I0479061

Essentials of Geography

Essentials of Geography

Anna Nelson

MURPHY & MOORE
www.murphy-moorepublishing.com

Essentials of Geography
Anna Nelson
ISBN: 978-1-63987-206-0 (Hardback)

© 2022 Murphy & Moore Publishing

MURPHY & MOORE

Published by Murphy & Moore Publishing,
1 Rockefeller Plaza,
New York City, NY 10020, USA

Cataloging-in-Publication Data

Essentials of geography / Anna Nelson.
        p. cm.
Includes bibliographical references and index.
ISBN 978-1-63987-206-0
1. Geography. 2. Earth sciences. I. Nelson, Anna.
G70 .E87 2022
910.01--dc23

This book contains information obtained from authentic and highly regarded sources. All chapters are published with permission under the Creative Commons Attribution Share Alike License or equivalent. A wide variety of references are listed. Permissions and sources are indicated; for detailed attributions, please refer to the permissions page. Reasonable efforts have been made to publish reliable data and information, but the authors, editors and publisher cannot assume any responsibility for the validity of all materials or the consequences of their use.

Trademark Notice: All trademarks used herein are the property of their respective owners. The use of any trademark in this text does not vest in the author or publisher any trademark ownership rights in such trademarks, nor does the use of such trademarks imply any affiliation with or endorsement of this book by such owners.

For more information regarding Murphy & Moore Publishing and its products, please visit the publisher's website www.murphy-moorepublishing.com

Table of Contents

Preface

This book aims to help a broader range of students by exploring a wide variety of significant topics related to this discipline. It will help students in achieving a higher level of understanding of the subject and excel in their respective fields. This book would not have been possible without the unwavered support of my senior professors who took out the time to provide me feedback and help me with the process. I would also like to thank my family for their patience and support.

The study of lands, inhabitants, features, physical properties and phenomena of the Earth and other planets is known as geography. It is considered as an all-encompassing field which seeks to understand the Earth and its natural and human complexities. Physical geography is the study of the Earth's landforms, soil, climate, seasons, streams, atmosphere and oceans. Human geography comprises of human, cultural, political, economic and social aspects. The branch of geography which describes the special aspects of interactions between individuals, society and their natural environment is called integrated geography. Four interrelated approaches are used for the study of geography. They are systematic, regional, descriptive and analytical approaches. A wide range of techniques are used within this discipline which include ethnographical research techniques, geostatistics, remote sensing, geographic information systems, etc. This book provides comprehensive insights into the field of geography. Such selected concepts that redefine this field have been presented in it. Those in search of information to further their knowledge will be greatly assisted by this book.

A brief overview of the book contents is provided below:

Chapter – What is Geography?

The field of science that deals with the study of lands, inhabitants, features and the phenomena of the Earth and other planets is called geography. It also studies the biological and cultural features of the Earth's surface. This is an introductory chapter which will briefly introduce about geography.

Chapter – Human Geography

Human geography is concerned with the study of people and their communities, cultures, economies, and interactions with the environment. It includes economic geography, historical geography, political geography, social geography, urban geography, etc. All these divisions under human geography have been carefully analyzed in this chapter.

Chapter – Physical Geography

Physical geography is the study of processes and patterns in the natural environment. Biogeography, geomorphology, climatology, coastal geography, oceanography, palaeogeography, etc. are studied under its domain. This chapter has been carefully written to provide an easy understanding of these branches within physical geography.

Chapter – Tools and Techniques used in Geography

There are various tools and techniques that are used in geography. It includes the use of geostatistics, cartography, geographic information system, geographic coordinate system, satellite imagery, geoinformatics, etc. This chapter closely examines these tools and techniques used in geography for a thorough understanding of the subject.

Chapter – Diverse Aspects of Geography

Some of the diverse aspects that fall under the field of geography are geoarchaeology, Tobler's first law of geography, glacial refugium, geographic contiguity, geo-literacy, geography of aging, etc. The topics elaborated in this chapter will help in gaining a better perspective of all the related aspects of geography.

Anna Nelson

1

What is Geography?

The field of science that deals with the study of lands, inhabitants, features and the phenomena of the Earth and other planets is called geography. It also studies the biological and cultural features of the Earth's surface. This is an introductory chapter which will briefly introduce about geography.

Geography is the study of the diverse environments, places, and spaces of Earth's surface and their interactions. It seeks to answer the questions of why things are as they are, where they are. The modern academic discipline of geography is rooted in ancient practice, concerned with the characteristics of places, in particular their natural environments and peoples, as well as the relations between the two. Its separate identity was first formulated and named some 2,000 years ago by the Greeks, whose geo and graphein were combined to mean "earth writing" or "earth description". However, what is now understood as geography was elaborated before then, in the Arab world and elsewhere. Ptolemy, author of one of the discipline's first books, Guide to Geography (2nd century ce), defined geography as "a representation in pictures of the whole known world together with the phenomena which are contained therein". This expresses what many still consider geography's essence—a description of the world using maps.

Camel near the Pyramids of Giza, Egypt.

To most people, geography means knowing where places are and what they are like. Discussion of an area's geography usually refers to its topography—its relief and drainage patterns and predominant vegetation, along with climate and weather patterns—together with human responses to that environment, as in agricultural, industrial, and other land uses and in settlement and urbanization patterns.

Although there was a much earlier teaching of what is now called geography, the academic discipline is largely a 20th-century creation, forming a bridge between the natural and social sciences. The history of geography is the history of thinking about the concepts of environments, places, and spaces. Its content covers an understanding of the physical reality we occupy and our transformations of environments into places that we find more comfortable to inhabit (although many such modifications often have negative long-term impacts). Geography provides insights into major contemporary issues, such as globalization and environmental change, as well as a detailed appreciation of local differences; changes in disciplinary interests and practices reflect those issues.

Shinjuku retail and entertainment district, Tokyo, Japan.

The history of geography has two main parts: the history of exploration and mapmaking and the development of the academic discipline.

The Emergence of Geography: Exploration and Mapping

As people travel, they encounter different environments and peoples. Such variations are intellectually stimulating: Why do people and places differ? Stores of knowledge were built up about such new and exotic places, as demonstrated by the Greek philosopher and world traveler Herodotus in the 5th century bce. That knowledge became known as geography, a term first used as the title of Eratosthenes of Cyrene's book Geographica in the 3rd century bce. Such was the volume of knowledge compiled thereafter that Strabo's Geography, published three centuries later, comprised 17 volumes. Its first two provided a wide-ranging review of previous writings, and the other 15 contained descriptions of particular parts of what was then the known world. Soon thereafter Ptolemy collated a large amount of information about the latitude and longitude of places in his seminal work.

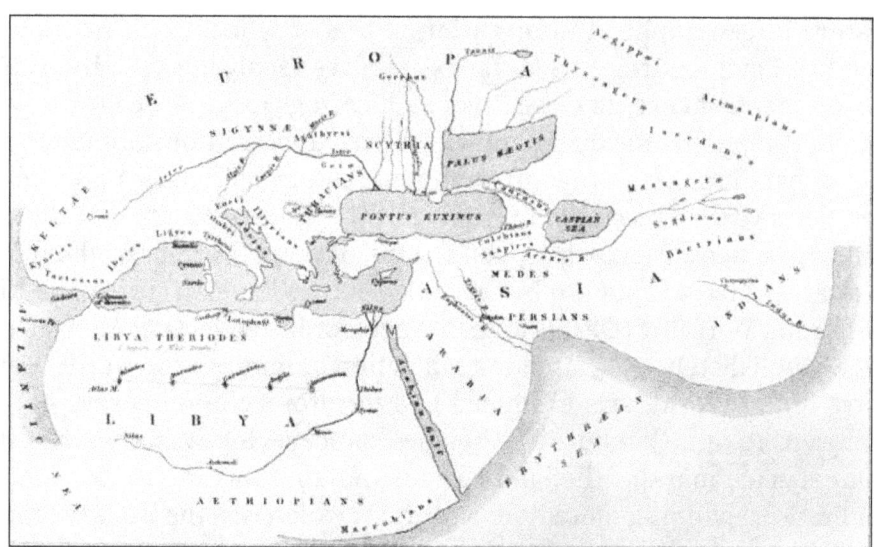

Map of the world, based on the description given by Herodotus.

The Greeks and Romans not only accumulated a great body of knowledge about Earth but also developed the sciences of astronomy and mapmaking, which helped them accurately locate places. However, during western Europe's Migration period (Dark Ages), much of that wisdom was lost, but the study of geography—notably cartography—was nurtured in the Arab world. This material became known to western Europeans during medieval times, partly through their contacts with the Muslim world during the Crusades. As the Europeans linked this new material with what. they could rediscover in ancient Greek and Roman work, they frequently stressed misinformation derived from the latter, notably in Ptolemy's inaccurate maps. From then on, as Europeans explored more of the world, increasing numbers of scholars collated new information and transmitted it to wider audiences.

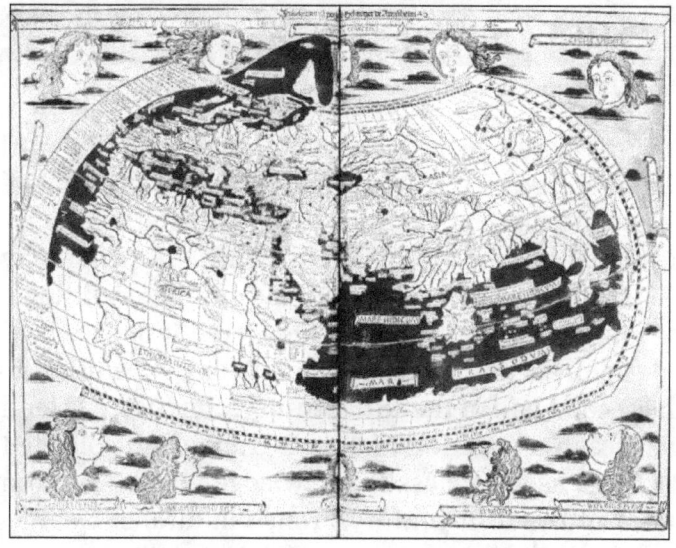

Ptolemy's map of the world, as printed at Ulm.

A key feature of geographical information is that it is localized, relating to individual parts of Earth's surface. Geography involves recording such information, in particular on maps—hence its close links with cartography. For centuries the locations of places were only inexactly known. Where to plot information on maps was frequently debated, as was drawing and demarcating boundaries around claimed territories. These debates were only resolved with more accurate and standardized cartographic practices. Meanwhile, collections of maps were assembled and published in atlases, a term first used by the 16th-century Flemish surveyor and cartographer Gerardus Mercator (Gerhard de Cremer) for his collection of maps of northern Europe, published in 1595; the first collection of maps of the world, Epitome of the Theatre of the World, was produced by Mercator's contemporary, the Belgian cartographer Abraham Ortelius. The science of surveying was employed to make detailed large-scale maps of the land surface; notable was the work of the Cassini family, in France, spanning more than a century, which was the basis for the world's first national atlas, published in 1791.

World map from Theatrum orbis terrarum ("Theatre of the World") by Abraham Ortelius.

Thus, the evolving practice of geography involved mapping the world, drawing outlines of what heretofore were terrae incognitae, and filling them in with details about their physical environments and the people inhabiting them. Such geographical advances depended on improvements not only in cartography but also in astronomy, which was vital for navigation. Methods for determining latitude and longitude and measuring elevations and distances were refined and were of great value to navigators and explorers and their sponsors. Many expeditions, such as those of James Cook in the second half of the 18th century, conducted scientific experiments that enabled advances in navigation and cartography and collected samples of flora and fauna that were used to classify knowledge about the natural world—as in the pioneering work of the 18th-century French naturalist Georges-Louis Leclerc, comte de Buffon. These links between geography, exploration, cartography, and astronomy have been maintained, appearing as the first sections of many contemporary atlases (with maps of the heavens along with terrestrial phenomena such as climate).

View of Huahine.

As information accumulated, a new branch of geography was established by the late Middle Ages, called chorography (or chorology). Books describing the then known world were used in geographical instruction at universities and elsewhere. Geography was not a separate discipline but was taught within established subjects such as mathematics and natural philosophy, in large part because it was of great importance to nation building and commerce. Among the early geography books were Nathaniel Carpenter's Geography Delineated Forth in Two Bookes and the German scholar Bernhardus Varenius's Geographia Generalis, which was revised and republished several times in the following century. Canadian geographer O.F.G. (George) Sitwell's catalog lists 993 different books in "special" (i.e., systematic) geography published between 1481 and 1887 in the English language; Lesley Cormack identified more than 550 geography books in the libraries of the Universities of Cambridge and Oxford in the period from 1580 to 1620.

Geography was practiced and taught largely because its information was valuable—notably for traders, those who invested in them, and the statesmen who supported both groups. By the early 19th century there was great demand for information and knowledge about the world. To aid commercial enterprises aimed at exploiting its resources and peoples, governments became involved in colonial ventures, annexing land beyond their frontiers, providing administrators and military protection, and encouraging settlement. All such endeavours required geographical information, including accurate maps. Increasingly, governments became directly involved in these activities, as with the U.S. government's sponsorship of major expeditions to the country's expanding western frontier and the establishment of national mapping agencies around the world.

Map of Lewis and Clark.

Geographical societies were established in many European and North American cities in the early 19th century to share and disseminate information. Among the first were those founded in Paris, Berlin London, St. Petersburg and New York City. Many of the European societies had royal patronage and strong support from the mercantile, diplomatic, and military classes. They collated and published information, sponsored expeditions, and held regular meetings, at which returning explorers might present their findings or participate in debates over technical issues such as mapping. These societies were central to the 19th-century mercantile and imperial ethos.

Geography and Education: The 19th-century Creation of an Academic Discipline

Geography's original characteristics were formulated by a small number of 19th-century French and German scholars, who strongly influenced subsequent developments in the United Kingdom and the United States. Since 1945, while retaining its focus on people, places, and environments, the discipline has expanded and changed considerably. Geography is one of the few academic disciplines, particularly in Europe, to have been established in universities as a result of pressure to produce people who could teach it in schools. As the demand for geographical information increased, more people required a foundation of geographical knowledge. There was also growing recognition of the role geography could play in creating national identities, making people aware of their particular situations through contrasts with environments and peoples elsewhere. Geographical knowledge was important to citizenship, especially if it supposedly showed the superiority of one's own people and environment.

Geography's links with mercantilism, imperialism, and citizenship were the basis of claims for geographical instruction in schools. For example, geographical societies lobbied successfully for their subject's inclusion in the curricula associated with universal school education, especially in northwestern Europe. Specialist bodies, such as the Geographical Association in the United Kingdom, continued to promote the discipline's educational role.

Sustaining the teaching of geography in schools required programs to train teachers and institutions where geographical knowledge could be codified and its scholarship advanced. Geography needed a presence in universities to give it academic credibility, and societies petitioned to secure it there. Some of this lobbying was successful by the end of the 19th century—the height of European imperialism. In Prussia, for example, a royal decree in 1875 established professorships of geography in 10 universities. In the Netherlands, the Royal Dutch Geographical Society was founded in 1873, largely to sponsor major expeditions to the Dutch East Indies. The society's first endowed chair, at a private university in Amsterdam, was specifically in "colonial geography". In Russia, St. Petersburg's Imperial Russian Geographical Society promoted the discipline in a variety of ways, establishing it early at Moscow State University. The Italian Geographical Society was founded in 1867, following the creation of the first university professorships in 1859; it too promoted "exploratory" geography and the teaching of geography in schools.

In the United Kingdom in the late 1880s, after such courses had been discontinued at the University of London, the Royal Geographical Society convinced Cambridge and Oxford to provide instruction in geography, with the society funding instruction for several decades (though degree courses were not introduced until the 1920s and '30s). As more British universities were founded, they too were pressed to provide instruction in geography. At some, private donations secured the appointment of lecturers. At others, a need for geography instruction was recognized in cognate disciplines, such as economics, geology, and history, although few of those appointed to do the teaching had any formal training in the discipline. This was also the case with the first professors of geography, appointed in the early 1930s at Cambridge and Oxford—Frank Debenham and Kenneth Mason, respectively.

Many of the first geography teachers were located in departments of disciplines that introduced geography teaching, but as the demand for courses grew—mainly from students who intended to teach the subject in schools—separate geography departments and degree programs were soon established. By 1945 there was a geography department in nearly every British university and in many of the universities and university colleges throughout the British Empire.

Geography's 19th-century research directions were set by a few influential individuals, although not all of them were even formally associated with the discipline. Many of its roots emanated from several continental European geographers, some of whom owed their inspiration to the teaching of philosophers such as Immanuel Kant, who wrote about geography in Critique of Pure Reason. Especially influential were the German scholars Alexander von Humboldt, Carl Ritter and Freidrich Ratzel and French geographer Paul Vidal de la Blache.

Humboldt's interests were stimulated by the Germans Johann Reinhold Forster and his son, Georg Forster, who on James Cook's second voyage had collated botanical and climatological data. Humboldt synthesized a vast amount of information (much of it

8

on his travels, including five years in Central and South America) to illustrate environ-mental variation, noting differences in agricultural practices and patterns of human settlement that reflected the interactions of elevation, temperature, and vegetation. His work emphasized field collection of data and their synthesis through maps, leading to inductive generalizations regarding environmental characteristics and their links with human activity. Materials assembled from a wide range of sources were the basis for his major published work, the five-volume Kosmos.

Whereas Humboldt laid the groundwork for what later became known as systematic geography, Ritter focused on regional geography, the study of the connections between phenomena in places. This involved defining regions, or separate areas with distinct as-semblages of phenomena. He relied on secondary data sources in compiling his 19-vol-ume Die Erdkunde im Verhältniss zur Natur und zur Geschichte des Menschen ("Earth Science in Relation to Nature and the History of Man"), which he never finished.

Ratzel, whose early studies were in biology and anthropology, was much influenced by Darwinian thinking when linking human societies to their physical environments. His two-volume Anthropogeographie related the course of history to Earth's physical fea-tures, illustrating the principle of survival of the fittest. His later Politische geographie utilized Darwinian arguments to characterize nation-states, which he treated as organ-isms that struggle for land (Lebensraum, or "living space"), only the strongest being able to expand territorially.

When geography was institutionalized at German universities in the late 19th centu-ry, however, there were no formally trained geographers, and the first professors had backgrounds in such disciplines as history, mathematics, geology, biology, or journal-ism. Their new discipline, which was conceived as a general Earth science, embraced systematic materials from those in which they had been trained. They created a unity for geography around the regional concept, building on foundations laid by Humboldt and Ritter. Field research outside Germany was deemed a crucial part of training, and each student spent a year overseas. Regular meetings involving several hundred Ger-man geographers—the Deutscher Geographentag—were held in the late 19th century, and these have continued to the present day. A separate Association of Academic Ge-ographers was formed in the early 20th century, by which time several important geo-graphical journals were in production.

In France the discipline had roots in history and mapping. The first major practitioner was Paul Vidal de la Blache, who had trained as a geographer and was appointed to the Sorbonne in 1898, where he maintained close links with the Annales school of histori-ans. Vidal focused on defining and describing regions, or what he called pays—relative-ly small homogeneous areas—whose distinctive genres de vie ("modes of life") resulted from the interactions of people with their physical milieux. Unlike some of his German contemporaries, notably Ratzel, he did not see those interactions as predominantly determined by the physical environment. Instead, he promoted what became known as possibilism, where the environment offers a range of options, and people choose how to modify nature according to their cultural and technological inheritances. As the contemporary historian Lucien Febvre put it, "nowhere necessities everywhere possibilities". Vidal's major contributions were his Tableau de la géographie de la France, an introduction to the multivolume Histoire de la France, and the 15-volume Géographie universelle. Many of his students wrote dissertations on individual pays, the study of which dominated French geography throughout the first half of the 20th century.

Developments in continental Europe during the late 19th century provided the foundation for an academic discipline to emerge in the English-speaking world, where the new intellectual concerns were integrated with the established traditions in exploration and cartography. The first International Geographical Congress was held in Antwerp, Belgium, in 1871; several more congresses were convened on the Continent before the first meetings in London and the United States and thus the notion was perpetuated, at least among some, that geography was still a "European" discipline. The International Geographical Union (IGU) was founded in 1922.

As a separate academic discipline, therefore, geography emerged out of a demand for teaching knowledge about the world's environments and peoples. From small and diverse beginnings, it was established in the academic community as a subject and developed associated institutions, such as learned societies to promote the discipline and journals in which geographers could publish their work, and its relevance grew to be recognized worldwide. In 1964, 70 countries sent delegates to the International Geographical Congress in London; now some 100 countries—through national committees for the discipline—are affiliated with the IGU. Geographers became academics in the full sense of the 20th-century university as they began to pursue research and original investigations.

In the United States, where each state was responsible for providing elementary, secondary, and higher education, there was no coordinated pressure for geography instruction. Instead, the creation of geography programs reflected particular local situations. In a number of universities, geography courses were offered for students in geology departments; in others, their origins lay in the universities' schools of business and commerce—as was the case at the University of California, Berkeley, where the country's first separate department of geography was created in 1898 within the College of Commerce. Out of these programs—and the courses included in teacher education at normal schools (many of which became state colleges and universities), as well as university education departments—emerged several geography departments, including graduate programs, such as those at the University of Chicago and at Clark University in Worcester, Massachusetts. By 1945 about 30 universities in the United States had geography graduate programs. However, the discipline remained relatively weak, and in subsequent decades several departments were closed and undergraduate degree programs discontinued. Notable were the closures at several Ivy League schools (the exception being Dartmouth College in Hanover, New Hampshire) and other private institutions, such as Northwestern University and, most conspicuously, the University of Chicago —although Harvard University (Cambridge, Massachusetts) established a Center for Geographic Analysis in 2006.

European scholars considerably influenced the emerging discipline in the United Kingdom and North America, where institutionalization into the academic structures came somewhat later. Scholars, some of whom studied in Germany or France, promoted different aspects of the discipline. Foremost in the United States was William Morris Davis, a geologist at Harvard University who published prolifically on landscape evolution (later called geomorphology, or the study of landforms). He argued strongly for education in geography, promoting an approach derived from German environmental determinism: human behaviour is strongly conditioned by environmental factors, so the study of physical geography should be the basis for understanding human activity. Davis was the principal author of an 1892 report on the teaching of geography, which recommended replacing the rote learning that characterized the discipline in American schools at that time with a more scientific approach based on physical geography but including "the physical influences by which man and the creatures of the Earth are so profoundly affected".

This approach was soon rejected as flawed by most geographers in the United States, who adopted a regional approach; areal variations in human activities, notably land uses, in their environmental settings were described, and homogeneous regions were defined. Richard Hartshorne codified this approach. His monograph, The Nature of Geography was much influenced by the work of German authors—notably Alfred Hettner—and it conceived the discipline's defining characteristics. Geography, he concluded, is:

> "A science that interprets the realities of areal differentiation of the world as they are found, not only in terms of the differences in certain things from place to place, but also in terms of the total combination of phenomena in each place, different from those at every other place."

Systematic geography focused on individual phenomena. But regional geography, or the study of the "total combination of phenomena" in places, was "the ultimate purpose of geography"—a task later redefined as "the highest form of the geographer's art". According to a leading British geographer, Sidney William Wooldridge, in The Geographer as Scientist: Essays on the Scope and Nature of Geography regional geography aimed:

> "To gather up the disparate strands of the systematic studies, the geographical aspects of other disciplines, into a coherent and focused unity, to see nature and nurture, physique and personality as closely related and interdependent elements in specific regions."

According to this view, all geographers—whatever their systematic interests in particular classes of phenomena—should be regional specialists who appreciate the full complexity of phenomena combinations. Many of Hartshorne's contemporaries identified themselves as regional geographers and published major texts, such as Preston E. James in his renowned Latin America. Many introductory texts, such as James's

An Outline of Geography used regional divisions of the world as organizing templates, though the regions were usually defined at much larger scales than the Vidalian pays.

Although regional geography dominated U.S. geographical practices in the first half of the 20th century, it was not universally adopted. Its major challenge was an approach—widely known as cultural geography—associated with Carl Sauer, a University of Chicago geography graduate, and the associates and students whom he led at the University of California, Berkeley, from 1923 to 1957. Sauer was also strongly influenced by the Germans, but he emphasized the study of landscape changes produced by the impress of different cultural groups on environments, with particular reference to rural Latin America. What became known as the Berkeley school used field, documentary, and other evidence to explore societal evolution in its environmental context, much of which apparently involved diffusion from core "culture areas".

These two approaches dominated U.S. geography for several decades, with considerable conflict between what were seen as the "Midwest" and "West Coast" definitions—which, respectively, were predominantly economic and cultural; Hartshorne's Perspective on the Nature of Geography partly reconciled the two, allowing for historical studies such as Sauer's. Not all American geographers followed one or the other, however. Some stressed systematic interests, as with early economic geographers such as J. Russell Smith, who worked in the Department of Geography and Industry at the University of Pennsylvania and published his Industrial and Commercial Geography in 1913. Economic or commercial geography courses were quite common in economics departments at American universities then, but with a shift in the focus of academic economics to a more analytical and statistical (or mathematical) approach, the links had nearly disappeared by 1920.

Also prominent in the early 20th century was Isaiah Bowman, president of Johns Hopkins University. A geology graduate of Harvard, where he was taught by William Morris Davis, Bowman did his early work on physical geography and pioneer settlement in South America. As director of the American Geographical Society he oversaw cartographic and other geographical work preparatory to the Paris Peace Conference which he attended, following World War I.

Two early influential geographers in the United Kingdom were both associated with the School of Geography at the University of Oxford. Halford John (later Sir Halford) Mackinder, appointed in 1887, was trained in the natural sciences and history and felt the need to prove his geographical credentials by climbing Mount Kenya in 1899. He is best known for his contributions to political geography; his concept of the "heartland"—the centre of the Eurasian landmass—as the pivotal area in world geopolitics influenced much Western political strategy for more than half a century. He later became a politician and diplomat. Mackinder actively promoted geographical education in schools. His 1887 paper to the Royal Geographical Society defined geography as the scientific study of the interrelationship between society and the environment. In addition, he convened the meeting in 1893 that founded the Geographical Association, which aimed to be a society for teachers of geography at all levels and became a successful lobby for the discipline.

Andrew John Herbertson took over the department at the University of Oxford after Mackinder. He drew on European roots and emphasized regional study, using climatic and other parameters to define regions at the global scale; others developed the regional concept, using a wider range of phenomena, at smaller scales (echoing the French work on pays). Regional geography remained at the core of the discipline in the United Kingdom until the 1950s, as promoted in The Spirit and Purpose of Geography by Sidney William Wooldridge and Gordon East.

Other influential early individuals included L. Dudley (later Sir Dudley) Stamp, a geologist by training who spent most of his career in the geography department of the London School of Economics. He directed a land-utilization survey of Britain in the 1930s, mobilizing some 250,000 students to map the country's land use. This material proved invaluable in agricultural planning during World War II, during which Stamp was involved in major government inquiries into land use, and was a foundation for his promotion of applied geography and geographers' contributions to the postwar extension of urban and rural planning activities. He also published many textbooks, stimulated interest in other areas (such as medical geography), and promoted collaboration through the IGU, of which he was president from 1960 until his death in 1966. Wooldridge was also trained as a geologist and worked in the King's College, London, geography department, where he was a major force in the development of physical geography in Britain—notably geomorphology, through his interpretations of Davis's ideas.

Another British geographer who influenced the discipline considerably through his own work and that of collaborators and graduate students was Henry Clifford (later Sir Clifford) Darby. The first to obtain a Ph.D. in geography at Cambridge, he pioneered work in historical geography through studies of landscape change and the detailed geography of England as displayed by the Domesday Book.Darby and his followers established a strong and continuing presence for historical geography early in the discipline's development in the United Kingdom.

Geography as a Science: A New Research Agenda

The then-established views regarding the nature of geography were set out in two large volumes in the early 1950s: Geography in the Twentieth Century edited by Griffith Taylor, and American Geography: Inventory and Prospect edited by Preston James and Clarence Jones. However, by then there was growing unease in North America and the United Kingdom with the dominant orientation of the discipline. It was seen as overemphasizing vertical (or society-environment) relationships and largely ignoring the horizontal (or spatial) relationships that characterized societies in which movement and exchange were so important. Geographers, it was argued, should pay more attention to spatial organization of economic, social, and political activities across the environmental backdrops. Too much effort was spent, as George Kimble expressed it:

> "Drawing boundaries that don't exist around areas that don't matter from the air it is the links in the landscape that impress the observer, not the boundaries."

Studies of areal functional organization were inaugurated, both for their intrinsic interest and because of their value; one pioneer, Robert Dickinson, argued that functional regions around towns and cities should be used to define regional and local government areas.

There was also a growing belief that the methods for defining regions were out of line with the scientific approaches characterizing other disciplines. Some felt that geographers had not contributed well to the war effort: Edward A. Ackerman, a professor of geography at the University of Chicago from 1948 to 1955 (and later head of the Carnegie Foundation), claimed that those working in the U.S. government's intelligence service had only a weak understanding of their material and portrayed them as "more or less amateurs in the subjects on which they published". He argued that geographers should follow not only the natural sciences but also most of the social sciences and should adopt more-rigorous research procedures.

Although there were moves in those directions in a number of places, the arguments were focused in 1953 by a paper in the prestigious Annals of the Association of American Geographers that strongly criticized what Ackerman called the "Hartshornian [i.e., regional] orthodoxy". Kurt Schaefer, a German-trained geographer at the University of Iowa, argued that science is characterized by its explanations. These involve laws, or generalized statements of observed regularities, that identify cause-and-effect relationships. According to Schaefer, "to explain the phenomena one has described means always to recognize them as instances of laws"; for him the major regularities that geographers study relate to spatial patterns (the horizontal relationships identified above), and so "geography has to be conceived as the science concerned with the formulation of the laws governing the spatial distribution of certain features on the earth's surface".

Schaefer codified what an increasing number of geographers were thinking, identifying a need for a major reorientation of—if not revolution in—its practices. The main thrusts occurred elsewhere. One of the most influential early centres was the University of Washington in Seattle, led by William Garrison and Edward Ullman. Their students, such as Brian Berry, William Bunge, Richard Morrill, and Waldo Tobler, became leading protagonists of the new geography, which rapidly spread to other universities in the United States, such as Northwestern, Chicago, and Ohio State in Columbus. It soon reached the United Kingdom, with initial centres at Cambridge and Bristol.

Much inspiration for these shifts came from economists, sociologists, and other social scientists, who were developing theories of spatial organization and using quantitative methods to test their hypotheses. The human geographers who followed their lead promoted in their practices what became known as the "quantitative and theoretical revolution". So too did physical geographers, who, for example, switched their focus from simply describing landforms to searching for scientific explanations of how they were created.

Three main arguments underpinned this paradigm shift in geographical practice. The first was that geography should become more scientifically rigorous, adopting the experimental science model (positivism) already in use by economists. The goal included deductive reasoning, which led to hypothesis testing with the goal of producing explanatory laws. The second was that such rigour required quantitative methods to provide precise descriptions and exact, reproducible research findings—unequivocal lawlike statements. Finally, with such a shift in disciplinary practices, the applied value of geographical work would be appreciated—in, for example, environmental and city and regional planning. Geography should be the science of spatial arrangements and environmental processes. Success in this promotion of geography as a science was crucial in winning recognition for the discipline in the United States from the National Science Foundation in the 1960s, initially as part of a Geography and Regional Science Program.

The success of those promoting change was assisted by the expansion of higher education. More students were going to colleges and universities, and new institutions were being founded. More geographers were needed to teach the subject, and many of those who were recruited preferred the novel approaches. The "revolutions" were to a considerable extent generational. The larger number of practicing geographers also precluded a small number of individuals imposing their views on the discipline; instead, there was encouragement to experiment and explore new topics and approaches. Furthermore, universities were increasingly emphasizing their research as well as teaching roles, and the new generations of geographers were more active as researchers than their predecessors. So more was done by more people, leading to greater specialization. Soon geography increasingly fragmented into specialist subdisciplines.

Physical Geography and Physical Systems

As a consequence of these changes, physical geography moved away from inductive accounts of environments and their origins and toward analysis of physical systems and processes. Interest in the physiography of the Earth's surface was replaced by research on how the environment works.

The clearest example of this shift came in geomorphology, which was by far the largest component of physical geography. The dominant model for several decades was developed and widely disseminated by William Morris Davis, who conceived an idealized normal cycle of erosion in temperate climatic regions involving the erosive power of running water. His followers used field and cartographic evidence to underpin accounts of how landscapes were formed: they constructed what geographers in the United Kingdom called "denudation chronologies". Davis recognized a number of other cycles outside temperate climatic areas in glaciated, desert, and periglacial and mountain areas, as well as in coastal and limestone areas. Each of these separate cycles had its own characteristic landforms. Because of long-term global climatic change, however, they may have characterized the now-temperate areas at different periods. For geomorphologists working in temperate regions, particular interest focused on the advance and retreat of glaciers during the Pleistocene Epoch. Landscape interpretation in many such areas involved identifying the influence of glaciations and the consequences of global warming, more recently a subject of considerable scientific interest. By the 1950s a major criticism of this work was that it was based on untested assumptions regarding landscape-forming processes. How does running water erode rocks? Only answering such questions could explain landform creation, and seeking those answers called for scientific measurement.

A series of photographs of the Grinnell Glacier taken from the summit of Mount Gould in Glacier National Park, Montana, in 1938, 1981, 1998, and 2006 (from left to right). In 1938 the Grinnell Glacier filled the entire area at the bottom of the image. By 2006 it had largely disappeared from this view.

There were three other main groups of physical geographers, two of whose work was also much influenced by the concepts of evolution. Workers in biogeography studied plants and, to a lesser extent, animals. The geography of plants reflects environmental conditions, especially climate and soils; biogeographical regions are characterized by those conditions and their floral assemblages, which produce patterns based on latitude and elevation. It was argued that those assemblages evolve toward climax communities. Whatever specific vegetation types initially occupy an area, competition between plants for available resources will lead to those most suited to the prevailing conditions eventually becoming dominant. Such conditions may change and a new cycle be initiated because of either short-term climatic fluctuations or human-induced environmental changes.

Wildflowers blooming on the tundra in Arctic National Wildlife Refuge.

The study of soils, or pedology, was concerned with the thin mantle of weathered material on the Earth's surface that sustains plant and animal life. World regions were identified based on underlying rocks and the operative physical and chemical weathering processes. Climatic conditions were important influences on soil types, with local variations reflecting differences in surface deposits and topography. As with landforms and plant communities, it was assumed that soils evolve toward a steady state, as weathering proceeds and characteristic soil profiles emerge for each region.

A gardener examining humus.

Finally, there was climatology, or the study of major world climatic systems and their associated local weather patterns in space and time. Much of the work was descriptive, identifying major climatic regions and relating them to solar and earth geometry. Others investigated the generation of seasonal and local weather patterns through the movements of weather systems, such as cyclones and anticyclones.

Hurricane Catarina, as viewed from the International Space Station.

These approaches dominated physical geography until the 1960s, when they were largely replaced. The new programs had three main aspects: greater emphasis on studying processes rather than outcomes, adoption of analytical procedures to measure and assess those processes and the associated forms, and integration of the processes into a focus on entire environmental systems. Many of the early changes involved detailed measurement of physical forms; deductive modeling based on physical properties developed later. Their integration into process-response models involved a reorientation of physical geography every bit as extensive as that in human geography. Physical geographers increasingly identified themselves as environmental scientists, using the basic concepts of physics, chemistry, and biology and the methods of mathematics to advance the understanding of how the environment works and how it produces its characteristic features.

The systems concept was a significant element of these changes. Climates, landforms, soils, and plant and animal ecology were conceived as being interrelated, with each having an impact on the other. The systems could be divided into subsystems with separate but linked characteristics and processes. Drainage basins became major units of study, for example, and were subdivided into the channels along which water is carried and the valley slopes whose form is created by the moving water. Geographers were introduced to the importance of studying systems by the work of a number of American geologists, such as Stanley Schumm and Arthur Strahler. However, the lack of interest in time and change—as expressed in Hartshorne's Nature—meant that little work had been done on physical geography in the United States for decades. The influential geographers included Briton Richard Chorley, who taught at the University of Cambridge after studying with Strahler in New York, and George Dury, who was trained in the United Kingdom but spent much of his career in Australia and the United States. These major protagonists introduced systems thinking and the study of processes to British physical geography, which was then reexported to American geography from the 1970s on, where locally trained individuals such as Melvin G. Marcus played key pioneering roles.

Salt-encrusted sand-lined drainage patterns at the fringe of the Etosha Pan.

Human Geography as Locational Analysis

In human geography, the new approach became known as "locational" or "spatial analysis" or, to some, "spatial science". It focused on spatial organization, and its key concepts were embedded into the functional region—the tributary area of a major node, whether a port, a market town, or a city shopping centre. Movements of people, messages, goods, and so on, were organized through such nodal centres. These were structured hierarchically, producing systems of places—cities, towns, villages, etc.—whose spatial arrangement followed fundamental principles. One of the most influential models for these principles was developed by German geographer Walter Christaller in the early 1930s, though it attracted little attention for two decades.

Christaller's central-place theory modeled settlement patterns in rural areas—the number and size of different places, their spacing, and the services they provided—according to principles of least-cost location. The assumption was that individuals want to minimize the time and cost involved in journeys to shops and offices, and thus the needed facilities should be both as close to their homes as possible and clustered together so that they can make as many purchases as possible in the same place. Likewise, businesses will want to maximize turnover, with people spending as much as possible on goods and services and as little as possible on transport. An efficient distribution of service centres was in the interest of both suppliers and consumers. Christaller showed that this required a hexagonal distribution of centres across a uniform plane (i.e., one that had no topographical barriers), with the smaller centres (providing fewer services) nested within the market areas of the larger.

Rural village.

Other works by non-geographers provided similar stimuli. Economists such as Edgar Hoover, August Lösch (who produced a theory similar to Christaller's), Tord Palander, and Alfred Weber suggested that manufacturing industries be located to minimize both input costs (including the costs of transporting raw materials to a plant) and distribution costs (getting the final goods to market). Least-cost location was the goal, which could be modeled as a form of spatial economics. Efficient spatial organization involved minimizing movement costs, which was represented by an adaptation of the physicists' classical gravity model. The amount of movement between two places should be a function of their size and the distance between them: i.e., size generates interaction, whereas distance attenuates it.

Chocolate factory.

These hypothesized patterns stimulated much searching for order in the distribution of economic activities and movements between places. Use of the intervening areas between the nodes and channels was also investigated within the same paradigm. A 19th-century German landowner-economist, Johann Heinrich von Thünen, had modeled the location of agricultural production, involving a zonal patterning of activities consistent with minimizing the costs of transporting outputs to markets with the highest-intensity activities closest to the nodes and channels. Economists adapted this to the organization of land uses within cities: these, and the associated land values, should be zonally organized, with housing density decreasing away from the centre and the major routes radiating from it.

Finally, there was the issue of change within such spatial systems, on which the work of Swedish geographer Torsten Hägerstrand was seminal. He added spatial components to sociological and economic models of the diffusion of information. According to Hägerstrand, the main centres of innovation tend to be the largest cities, from which new ideas and practices spread down the urban hierarchies and across the intervening nonurban spaces according to the least-cost principles of distance-decay models. Later studies validated his model, with the best examples provided by the spread of infectious and contagious diseases.

The models of patterns and flows were synthesized to describe urban systems at two main scales: systems of cities, in which places were depicted as nodes in the organizational template, and cities as systems, focusing on their internal organization. The goal was not just to describe those systems and their operations but also to model them (statistically and mathematically), thus producing general knowledge about the spatial organization of society rather than just specific knowledge about individual places. Location-allocation models suggested both optimum locations for facilities and efficient flows between them. A new discipline, regional science, was launched by economist Walter Isard to study such systems and promote the application of the knowledge acquired. It failed to gain separate status within universities, but many geographers still participate in its conferences and publish in its journals.

By the late 1960s these new practices were synthesized in influential innovative textbooks on both sides of the North Atlantic. Notable examples include Peter Haggett's Locational Analysis in Human Geography Richard Chorley and Haggett's Models in Geography, Ron Abler, John Adams, and Peter Gould's Spatial Organization, and Richard L. Morrill's The Spatial Organization of Society. Each emphasized the theme earlier pronounced by Wreford Watson that "geography is a discipline in distance".

The early models made relatively simple assumptions regarding human behaviour; the principle of least effort predominated, with monetary considerations preeminent, and it was assumed that decisions were based on complete information. These were later relaxed, and more-realistic models of spatial behaviour were based on observed decision making in which the acquisition and use of information in spatial contexts took centre stage. Distance was one constraint on behaviour; it was not absolute, however, but manipulable, as patterns of accessibility could be changed. And as the behavioral contexts were altered, the learning and decision-making processes within them also changed, and the spatial organization of society was continually restructured.

As research practices changed, so too did teaching. The earlier focus on field observation, map interpretation, and regional definition was replaced, and research methods for collecting and analyzing data—particularly statistical analysis—became compulsory elements in degree programs. New subdisciplines—notably urban geography—came rapidly to the fore, as systematic specialisms displaced regional courses from the core of many curricula. Other parts of the discipline—economic, social, political, and historical—were influenced by the theoretical and quantitative revolutions. What became known as a "new" human geography was created, initially at a few institutions in the United States and the United Kingdom but rapidly spread through the other Anglophone countries and later to a variety of other countries.

Methods and Machines

Mapmaking and Remote Sensing

The map was long the geographer's main tool, with map construction and interpretation being the major practical skills taught in degree programs. Mapmaking involved knowledge of surveying and projections, in addition to the arts of depicting point, line, and area data on maps. Map interpretation involved their use not only in the field for location but also in the laboratory for identifying landscape and other features, with map comparison used to identify associations among distributions and to define regions with multiple criteria. Alongside the map—especially after World War II—geographers increasingly used aerial photography to supplement these landscape-interpretation skills.

Aerial view of the Amazon rainforest.

By the end of the 20th century, very little of this material remained in degree curricula; mapping skills were seldom a significant part of the geography student's education. Mapmaking was moved from the field and drawing board to the laboratory and keyboard, using remotely sensed imagery, geographical positioning systems (e.g., the Global Positioning System [GPS]), and computers. So was the production of maps to display patterns of interest to geographers; standard computer software packages provided geographers with their illustrative material without any need to use pen and ink.

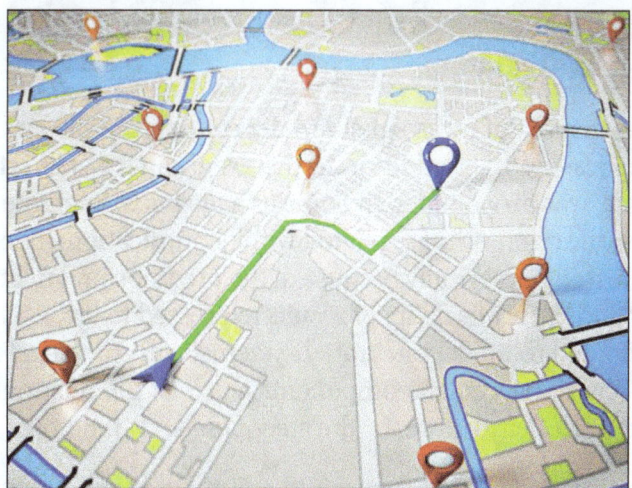

GPS-derived navigation map.

The analysis of remotely sensed images—initially from airplanes but increasingly from spacecraft—assumed considerable importance in some areas of geographical research, especially physical geography. Images provided immediate, regular, and frequent information on parts of the world that were difficult to access physically, making it possible not only to produce detailed maps but also to make estimations of environmental conditions (such as biomass volume, soil wetness, and river sediment loads) and to assess short-term changes. Such images are the only source of data at the global scale and are increasingly important for modeling environmental changes.

A Global Positioning System (GPS) satellite orbiting Earth.

Much experimentation was required to realize the potential uses of the massive volume of data provided from spacecraft sensors, and remote-sensing techniques became important tools; radar, for example, circumvented the problem of generating images in cloudy areas. The techniques for producing these newer images were largely the province of physics, mathematics, and computer science. Geographers were concerned with their use in understanding and managing the environment, with field studies providing the ground data against which image assessments could be evaluated, and developing remote-sensing methods for various tasks, such as estimating precipitation in desert areas.

Computational Analysis

The use of remote-sensing data was substantially confined to physical geographers, but the use of mathematics—another addition to the geographers' skill sets—was used more widely and, for a time, was propounded by some as a means to integrate human and physical geography. Scientific rigour was associated with quantification; identities and relationships had to be expressed numerically because of the precision and unambiguity of mathematical statements and the replicability of results expressed in those terms. Mathematical procedures were adopted to model integrated systems, with statistical methods deployed to test hypotheses regarding system components, such as the relationship between land values and distance from a city centre, or the steepness and stability of a range of slopes.

Geographers initially assumed that they could adapt standard statistical procedures to their particular problems, exploring the validity and viability of a range of approaches (from econometrics, biometrics, psychometrics, and sociometrics). The greatest emphasis in these pioneering applications and textbooks was placed on methods associated with the general linear model—e.g., regression, correlation, analysis of variance, and factor analysis—but specific spatial statistical procedures for analyzing point and line patterns were also explored.

Geographers soon realized that spatial data present specific analytical problems that require particular treatment and for which standard procedures have to be modified. A wide range of issues in geostatistics was identified, such as the problems of spatial autocorrelation in analyzing all spatial data, the modifiable areal unit problem and associated ecological fallacies in human geography, and the means of estimating values on maps from what is known about neighbouring sites. Analyzing spatial data has been enormously facilitated by developments in computer power and algorithms. Advancements in computational skills have allowed geographers to not only address previously intractable problems but also provide a means for thinking about problems that were not even considered before technology enabled them.

Geographic Information Systems

The major technological advance of the late 20th century in this regard was one that, although not specific to geography in its wide range of applications, has had particular resonance for geographers. Geographic information systems (GIS) are combined hardware and software systems for the capture, storage, checking, integration, manipulation, display, and analysis of spatially referenced (geocoded) data. The data (i.e., information with coordinate referencing, such as latitude and longitude) are input into these systems and displayed in two- or three-dimensional maps and other diagrammatic forms. Two or more maps can be overlaid and integrated for analysis—such as a relief map and a map of wells—even if they are compiled on different spatial grids. If geocoding schemes can be made compatible, separate data sets can be combined, analyzed, and displayed. This is technically demanding in many circumstances because of the issues involved in the interpolation of values for particular points from partial data. GIS facilitates modeling of processes in both space and time and has been the focus of much research investment. It has a massive range of potential applications in a wide range of areas, such as the planning of public facilities and services.

The development of GIS and their applicability is a significant focus of contemporary geographical work. Major public initiatives in the late 1980s in both the United States and the United Kingdom—the National Center for Geographic Science and the Regional Research Laboratories, respectively—have allowed research to expand considerably, with geographers at the centre of major components of the information sector (i.e., those who produce and disseminate information). Instruction in GIS operation and use is now a core component of many degree programs. Many universities offer specialist qualifications in GIS, and conferences of GIS users are by far the largest regular gatherings involving geographers. To some this modern expression of cartography comprises a geographic information science, part of a larger field of geoinformatics; it provides many geography graduates with a heavily demanded key skill, and its research and applications potential offers a secure foundation for the discipline's future.

Growth, Depth and Fragmentation in the Late 20th Century

Once the switch from inductive reasoning based on field evidence to deductive modeling and field-testing had been generally accepted within physical geography, change in that section of the discipline became more gradual and progressive rather than punctuated by significant advances. The last decades of the 20th century were marked by greater sophistication in modeling, data collection, and analysis—by a deepening of the discipline and a greater integration of its parts. Increasingly, physical geographers identified themselves as earth systems scientists, and their peer group became practitioners in a wide range of sciences, rather than other (especially, but not only, human) geographers. Physical geographers have retained distinctiveness in this wider enterprise through their abilities at handling spatial data and the problems of collecting and analyzing field data—skills increasingly deployed in large multidisciplinary projects.

Such continuity was not so readily apparent in human geography, whose practitioners have generated almost constant debate over its nature and methods without any one approach becoming dominant. As a result, human geography has become more fragmented than physical geography. This has been facilitated by continued growth in the number of practicing geographers, especially in the United Kingdom, where the discipline's popularity and strength in the universities has ensured the needed resources.

Influence of the Social Sciences

New practices in human geography have been closely linked to parallel changes in the social sciences, in some of which the quantitative-positivist approach has come under attack. The arguments were extended to the spatial-analysis approach with its geometric emphasis. By reducing all decision making to economic criteria, subject to immutable laws regarding least costs, profit maximization, and distance minimizing, geographers, it was claimed, were ignoring (even denigrating) the role of culture and individuality in human behaviour. By proposing to use those laws as bases for spatial planning, they were simply reproducing the status quo of capitalist domination; and by assuming universal patterns of behaviour, it was argued, they were patronizing those who chose to operate differently.

Stimulating and growing out of these arguments were three main strands of work. In the first, geographers led by David Harvey (who was Cambridge-trained but worked largely in the United States) explored Marxist thinking. This involved not only the workings of the economy—to which they added an important spatial dimension—but also the class conflict underpinning Marxian analyses and the consequent unequal distribution of power. The positivist aspects of locational analysis were attacked as largely irrelevant; they assumed constant conditions for economic decision making and, thus, universal laws of behaviour, whereas for Marxist scholars continuous change was the norm.

A popular alternative approach for some of a generally Marxist persuasion was critical realism. This accepts that there are general tendencies within capitalism but contends that they are only realized when implemented by individuals making decisions in local contexts: the profit motive is general, but individual entrepreneurs decide how to pursue it. The outcomes then change the local contexts—for example, by changing the maps of economic activity within which decisions are made, so that the contingent circumstances for future decisions also change—and there can be no general laws of outcomes, only of basic processes. This argument was forcefully made by the British geographer Doreen Massey. Furthermore, decision makers learn from the consequences of previous decisions. There is a continuous interplay between context and decision maker (or between structure and agency). Realists can explain why events have occurred—why a factory is located at a particular site—but not as examples of general laws of location. For them, explanation means accounting for specific events in context, relating how decision makers react to circumstances in order to meet imperatives within the constraints of their particular situations (what they know, what they believe their competitors will do, and how they manipulate that knowledge).

Marxist-inspired approaches to understanding spatial arrangements covered a wide range of issues, many relating to inequalities in society. Access to various goods and services—e.g., housing and health care—is a function of class position, not only locally and regionally but also nationally and internationally. The geography of development, embracing not only wealth and income but also the quality of life and life chances, reflects a global economic system that varies at several levels.

Marxism is more than a mode of analysis based on axioms regarding capitalist economic systems: it has an associated politics. Many geographers inspired by this approach in the context of the world situation in the 1960s and '70s were attracted to the politics and adopted the term "radical geography". Others accepted the power of Marxist-inspired analysis without also agreeing with the associated socialist agenda. From these twin positions, a more broadly based critical geography emerged that identified spatial problems of contemporary societies and their causes and promoted solutions, while at the same time meeting principles of social justice and ethical practice.

This critical geography also drew on a second strand of work, which developed out of writings on gender and the growth of feminist scholarship. Feminist geographers contended that geography was a male-dominated discipline whose concerns reflected masculinist epistemologies. Women were subordinated and largely ignored in geography, and feminists pointed out the gender divisions and campaigned to remove bias against women. In spatial science, for example, they showed how patterns of accessibility discriminated against women in labour markets, demonstrating how space had been manipulated to promote male interests and in the process had become part of society's definition of gender roles.

Feminists also contended that gender is one of the multiple positions that individuals occupy within a society, rejecting the predominant class position at the core of Marxian analyses. From this foundation emerged wider concerns with identity and positionality, embracing not only gender divisions but also ethnic and national distinctions, as well as sexual orientation and other criteria on which individuals' identities are based—such as the position of those in postcolonial societies. Thus, gender had to be subdivided to recognize the different positions (and politics) of white and black women, of women in societies with developed and developing economies, and in various religions. Appreciating those divisions—plus the many hybrid positions that emerge through, for example, the mixing of peoples in multiethnic cities—requires appreciating discrimination and difference. To many, this cannot be achieved by the abstract theorizing of either spatial science or Marxian analysis. It requires interpretative methodologies aimed at understanding through empathy, gained through a variety of qualitative research methods, such as participant observation, focus groups, in-depth interviewing, and the examination of archived resources. These enable access to not only how people interpret their place in the world and act accordingly but also to how they create worlds within which to act, at all spatial levels from the smallest (their individual bodies) outward.

An example of such analyses is critical geopolitics. Political geography was a marginal subdiscipline for several decades after World War II, with geopolitical thinking disparaged because of its association with the work of geographers in 1930s Nazi Germany. Its revival involved regaining an appreciation of how influential political thinkers and politicians develop and propagate mental maps of the world as structures for action. These mental maps are created by key thinkers, adopted by politicians, and disseminated by various media. They form contexts for developing political strategies and determining tactics, to which the wider population's attitudes are molded. The world of politics is a world of mental maps and of dominant views that underpin behaviour: we act in perceived worlds that intersect with, but are often more powerful than, real worlds, which are composed of physical phenomena.

Such work came to be associated with another major development in the social sciences: postmodernism. This concept maintains that there are no absolute truths, so no grand theories can provide universal explanations and guides to action. Truths are the beliefs on which people act, and there are multiple truths of which none can claim primacy, although the value of competing truths in any context can be assessed ethically, according to local conceptions of right and wrong. People learn their truths from others—through either direct or indirect sources. Therefore, much learning takes place in contexts, and, since most people live relatively spatially constrained lives, those contexts are territorially defined. They are the places and areas within which people interact and learn—their homes and neighbourhoods, their schools and universities, their workplaces, and the formal organizations in which they participate—and that they create and maintain through local interactions.

This appreciation of the role of context put the concept of place on centre stage in much human geographical research, displacing space from the primary position it occupied for several decades. It differs from the former regional tradition in which environmental features dominated. Places are defined more fluidly: they are made, remade, and dissolved by people; they may overlap, or they may be bounded and defended. Places occupy core positions in human existence and everyday lives. People learn attitudes and behaviour patterns in places where they interact with others and to which they ascribe meanings—a theme developed by humanistic geographers over several decades, as in books on topics such as Topophilia: A Study of Environmental Perceptions, Attitudes, and Values, by Yi-fu Tuan. Their identities and their politics are associated with the nature of their places. As people learn and change themselves, so too do they change their environments. Furthermore, as critical geopolitics illustrates, such place making involves not only creating an identity for one's home area but also separate identities for those of other areas. Geographers have been stimulated by Edward Said's Orientalism, which portrays how Western societies created images of the East in opposition to themselves. These images, portrayed in literature and other media, are the basis for attitudes toward many non-Western cultures, presenting "the other" as not only different but also inferior and thus not deserving equal treatment and respect—as was exemplified in Derek Gregory's seminal The Colonial Present: Afghanistan, Palestine, Iraq.

This revived interest in places is a feature of the third contemporary strand, with geographers engaged in the field of cultural studies, which encompasses scholars from the humanities and social sciences studying human action in context. Such work ranges over many aspects of behaviour, including the microscale of the individual body, and seeks to understand the meanings that underpin actions—many of which are never recorded during the processes of everyday life—and how communities and groups identify with places and spaces. The relationships between people and nature are also being reconsidered, breaking down artificial boundaries between these long-considered opposites. New approaches for interrogating actions are being explored: geography quite literally studies where events take place, and the impact of those events is reflected in places' characters. Indeed, such is the contribution of geography to cultural studies that some identify a "spatial turn" within the humanities.

Linking the Human and Physical Worlds

There has also been an increasing stream of work on the interactions between human societies and physical environments—long a central concern for some geographers, as illustrated by Clarence Glacken's magisterial treatment of Western interpretations of nature in Traces on the Rhodian Shore: Nature and Culture in Western Thought from Ancient Times to the End of the Eighteenth Century. Human abuse and despoliation of the environment are important themes introduced in their modern context by a pioneering American conservationist, George Perkins Marsh, in Man and Nature, but they were minor concerns among most geographers until the late 20th century.

People walking through a flooded street.

One significant example of work on the interaction of human society and nature was stimulated by Gilbert White, a geography graduate of the University of Chicago. White returned to Chicago in the 1950s to lead a major research program on floodplains and their management, assessing people's views of the risks of floodplain use and evaluating the influence of flood insurance on their actions. From that foundation, White and his coworkers pioneered research into a wide range of environmental hazards and risk taking and the development of sustainable environmental management strategies, and they were also involved in government and international agency programs.

When environmental concerns moved to centre stage politically and publicly in the 1970s, relatively few geographers were working on society-nature interrelationships; topics that they considered within their discipline's purview were being commandeered by biologists, earth scientists, and sociologists, for example, and new subject areas such as environmental history. Over time, four main themes—environmental influences on human activities, the impact of humans on environmental processes, environmental conservation, and environmental management—formed a growing corpus of geographical work on environmental issues. One area of interest has been environmental attitudes and ideologies and environmental meanings and understandings within different societies. Others have studied environmental politics, environmentalism as a basis for political action, environmental policy making, policy assessment (as with environmental risk analyses), and the role and interpretation of environmental risks and hazards in human decision making.

Environmentalists in front of the Eiffel Tower during the Paris Climate Change Conference, their bodies positioned to create a message delivering their sentiments about renewable energy.

Some see the environmental focus as a means not only of establishing the relevance of geography to pressing public concerns but also of reintegrating physical and human geography. There has been some coming together but little close engagement between the two subdisciplines, largely because they define knowledge quite differently. The scientific foundations of modern physical geography sit uneasily with the qualitative and critical research methods of many human geographers. Nevertheless, interest in the society-nature nexus has increased and has given the discipline a clear identity within the sciences.

The Contemporary Discipline

The academic discipline of geography is extremely broad in subject matter and approaches; it contains specialists covering diverse subjects but sharing concerns over places, spaces, and environments. Indeed, the discipline is now fragmented into a substantial number of separate subcommunities among some of which there is relatively little contact. The Association of American Geographers has more than 50 separate specialty groups, for example, catering to its members' particular interests. Some physical geographers have stronger links outside their discipline than within it. The International Geographical Union—based in Rome at the "Home of Geography," provided by the Italian Geographical Society—has some two dozen commissions and about a dozen study groups.

Given this diversity of interests, encapsulating the contemporary discipline in only a small number of categories is difficult. The main division continues to be between physical and human geography, each of which contains subdivisions and even sub-subdivisions.

Physical Geography

Since the reorientation after 1970 of physical geography to the study of systems of natural environmental processes, there have been major changes in both research and teaching. Much research now involves large, tightly focused collaborative programs of careful measurement, modeling, and analysis. It is much more demanding and expensive in resources than previously: equipping field expeditions and laboratories and learning related techniques necessarily generates specialization. This is facilitated and integrated by major international interdisciplinary programs, such as those associated with the United Nations Educational, Scientific, and Cultural Organization (UNESCO) and the European Union (EU), as well as national research councils and major government research bodies such as National Aeronautics and Space Administration (NASA). Typical of this shift has been the relative demise of the study of landforms. There are now two main research communities within geomorphology: those who study contemporary processes and those who investigate environmental change and landscape evolution since the beginning of the Quarternary Period (about 1.8 million years ago).

The importance of water in erosion plus the transport and deposition of sedimentary materials is reflected by work in geographical hydrology. This relative emphasis on water in contemporary physical geography undoubtedly indicates the concentration of English-speaking geographers working in temperate latitudes. There is also substantial work in glaciology, reflecting ice's role in creating many current temperate environments, as well as—especially in the case of polar ice—in contemporary climatic change. Similarly, much work is being done on dryland areas, a consequence of political as well as intellectual interest in desertification and land degradation.

Melting glacial ice.

Other areas of the natural environment attract less attention. There are few large re-search teams in biogeography. Remotely sensed data are used to map land cover, how-ever, to estimate biomass and model ecosystems for work on biodiversity and the car-bon cycle, and to chart disturbances generated both naturally and by human-induced events (e.g., bushfires). The geography of soils is only a minor field of study, with some work on erosion and reclamation. Advances in climatology involve extremely large-scale computer modeling from global to local focus, based on understanding atmospheric physics and meteorology; relatively little of this involves geographers, whose main con-tributions concern physical, synoptic, and applied climatology and climatic impacts (i.e., on agriculture). These three subdisciplines remain part of many geography degree programs, however; indeed, geography departments offer more introductory work on various aspects of the environment, at all scales, than do most other sciences.

Physical geography now concentrates on the Earth's surface processes, therefore, in-volving field and laboratory investigations of contemporary processes and the re-construction of past environments, especially the relatively recent past (which in-cludes collaboration with archaeologists). These are integrated in research programs into past, contemporary, and future environmental changes. Concern about global warming and climate change, sea-level changes, extreme environmental events, and the loss of biodiversity stimulated modeling of environmental systems involving the interactions among the Earth's hydrological, ecological, and atmospheric components. Building large models of these systems and their complex interrelationships involves teams seeking not only to understand their operations but also to predict environmen-tal futures as bases for public policy making at global, international, national, and local scales. Research reconstructing past environments puts current processes and changes into longer-term perspective.

The methods employed by physical geographers are those of environmental scientists more generally; knowledge of relevant work in physics, chemistry, biology, and mathe-matics is necessary, and applications increasingly involve working with engineers. Ge-ographers have developed particular areas of expertise within environmental science, as with the analysis of remotely sensed data. Processing the massive databases produced daily involves major geocomputation expertise to address these questions: What is where? How much is there? What condition is it in?

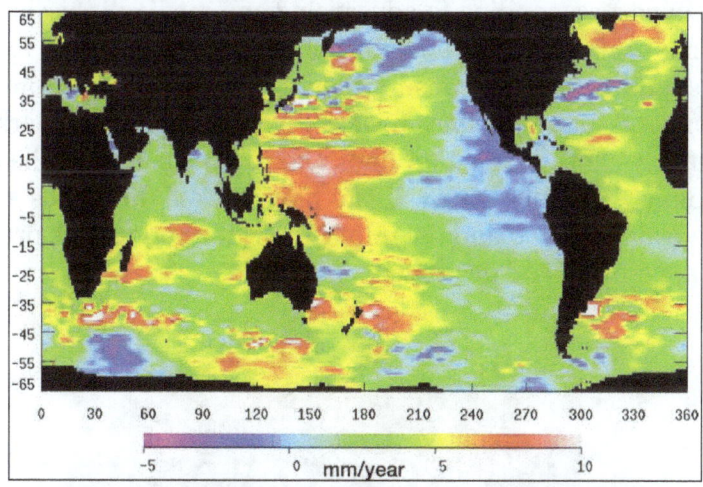

Change in sea surface height from 1993 to 2008, using data collected from the Topex/Poseidon and Jason-1 satellites.

Human Geography

Since 1945 human geography has contained five main divisions. The first four—economic, social, cultural, and political—reflect both the main areas of contemporary life and the social science disciplines with which geographers interact (i.e., economics, sociology, anthropology, and political science and international relations, respectively); the fifth is historical geography. All five have remained central, being joined in the mid- to late 20th century by concentrations on particular types of areas, notably urban. Research interests in specific regions have declined, and relatively few geographers now identify themselves as experts on a particular part of the world.

Economic geography has a long pedigree. Its traditional focus has been the distribution of various productive activities—with subdivisions into, for example, the geography of agriculture, industrial geography, and the geography of services—and patterns of trade such as transport geography. Such concentrations were strengthened by the move into spatial analysis. Relatively little work in that mold is now undertaken, however, and the models of idealized economic landscapes that dominated in the 1960s and '70s are now rarely deployed or taught. Part of the change reflects economic shifts, notably the extension of globalization. Transport costs have decreasing significance for many location decisions, relative to labour and other costs. Instead, the decision making of transnational corporations dominates the changing global pattern of activity, reflecting a wide range of political as well as economic concerns regarding the profitability of investing in different countries and regions. Much contemporary work studies company locational decision-making processes, the regulatory regimes of individual states (including policies designed to attract and retain investment), and their impact on the pattern of economic activity.

Economic and cultural worlds are closely intertwined. Many individual economic decisions in advanced industrial countries—e.g., what to buy, where to eat, and where to take vacations—reflect not needs but rather culturally induced preferences, which change rapidly, in part responding to advertising and media discussions of tastes and fashions. To some commentators, this generates a significant shift in the major features of capitalist production and consumption. It is moving away from mass products manufactured on large assembly lines toward myriad small niche markets with factories having relatively short production lines and rapid changes in the details of their products. Economic geographers investigate how markets for goods and services are culturally created and changed and the implications for both where production occurs and where jobs are created and destroyed.

Resident casting his vote in a local election.

Political geography also has a considerable pedigree, although it attracted little attention during the mid-20th century. Its main concerns are with the state and its territory—with states' external relations and the relationships between governments and citizens. The geography of conflict incorporates both local conflicts, over such matters as land use and environmental issues, and international conflicts, including the growth of nationalism and the creation of new states. Electoral geography is a small subfield, concerned with voting patterns and the translation of votes into legislative seats through the deployment of territorially defined electoral districts.

Social geography concentrates on divisions within society, initially class, ethnicity, and, to a lesser extent, religion; however, more recently others have been added, such as gender, sexual orientation, and age. Mapping where different groups are concentrated is a common activity, especially within urban areas, as is investigating the related inequalities and conflicts. Such mappings are complemented by more-detailed studies of the role of place and space in social behaviour—as with studies of the geography of crime and of educational provision—and in how mental representations of those geographies are created and transmitted.

Other subdisciplines associated with social geography are sometimes seen as separate. Population geography is largely concerned with the three main demographic characteristics of fertility, mortality, and migration; investigations using census and other data are complemented by detailed case studies of decision making, such as whether and where to migrate and how relevant information is received and processed. Medical geography focuses on patterns of disease and death—of how diseases spread, for example, and how variations in morbidity and mortality rates reflect local environments—and on geographies of health care provision.

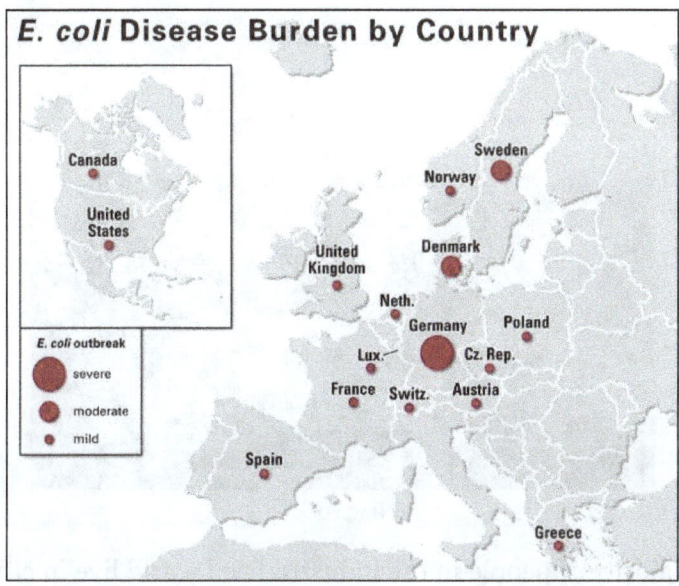

E. coli Disease Burden by Country

The severity of the E. coli disease outbreak of 2011 shown by country.

In its original manifestations, cultural geography had close links with anthropology, especially in the work of Sauer's Berkeley school. This has been superseded by a wider appreciation of the interrelationships among people and societies as well as between people and their environments. Cultures are sets of beliefs transmitted in various ways. Many involve texts, not only written but also visual and constructed (e.g., works of art and architecture) and aural (e.g., soundscapes); some may never be recorded but are transitory moments in people's movements and expressions. Interpreting them involves deconstructing what people say and do, activities that bring geographers into contact with the humanities as well as the social sciences in developing appreciations of meanings in texts and actions, including the landscapes and townscapes—large and small, personal and intimate, as well as grand and public—created in the process.

Places are central to this diverse range of contemporary work, especially in the study of cultural change, which involves mixing people from different backgrounds and areas as they move through space. Cultures are fluid and continually renegotiated, as are the spaces they create and occupy. Many negotiations involve conflict and the exercise of power—for example, the imperial strategies in the construction of 19th- and 20th-century worlds and postcolonial responses to others' worldviews imposed on them.

One of the most popular fields of study from the 1960s to the '80s was urban geography, under the banner of which much pioneering work in the locational analysis approach was conducted. Cities and towns were field laboratories for testing models of least-cost decision making. When those models were dismissed as oversimplifying complex realities and the search for spatial laws about cities declined, interest turned to contemporary concerns regarding urban areas and life. Cities are major globalization nodes, economic power being centralized in a small number of world cities (London, New York City, and Tokyo are usually placed at the top of city hierarchies).

New York.

Given that the majority of people in the industrialized world live in cities, it is not surprising that urban geography has received much more attention than rural geography. Relatively little work was done on aspects of rural areas other than agriculture before the 1970s, just when, according to some, much of the particularity of rural areas was disappearing as many features of urban society were reaching into the countryside. To others, however, issues unique to rural, low-density areas call for a separate rural geography; although typical urban problems such as poverty, homelessness, social exclusion, and access to public facilities are also characteristic of rural low-density areas, particular issues there include the society-nature relationships, common images of the "rural," and the role of tourism in reinvigorating rural economies.

Historical geography has retained its identity and distinction, although historical geographers have not distanced themselves from changes elsewhere in the discipline, with which their focus on interpreting the past from available evidence resonates. The developments in locational analysis stimulated some new ways to study available data. For others, the later developments, especially in cultural geography, coincided with their deployment of a wide range of nonquantitative sources to reconstruct the real and imagined, as well as the abstract (spatial analysis), worlds of the past; issues of postcolonialism have attracted the attention of historical geographers as well as those interested in current cultural issues.

A great range of sources is now used in such endeavours, not only maps but also, for example, travelers' writings about worlds they have encountered. Within this enterprise is a rejuvenated interest in the history of geography itself, not merely as a means of better appreciating where the discipline has come from but also of illustrating the importance of place and context in its evolution; geography, like so much else, is a range of practices that emerged and evolved in response to local stimuli. Geographers have produced particular forms of knowledge that have been significantly influenced by how people have encountered the world.

People and the Environment: The Physical and the Human

Historical geographers have long investigated landscape change. Their work now informs investigations of global environmental changes as well as illustrating past human-induced environmental modifications. Other research evaluates contemporary environmental changes and their implications not only for environmental futures but also for individual life chances.

Such studies occupy the intersection of physical and human geography, although relatively little work involves collaboration among human and physical geographers. For the latter, it involves incorporating human-induced changes to models of environmental processes and systems. Human geographers' concerns range widely, from pragmatically applied work on environmental policy and management through political ecology to explorations of culture-nature interrelations.

Methods of Geography

Changes in what a discipline studies are closely interwoven with changes in how its research is undertaken. Some substantive changes have been technologically driven: without new facilities, advances would not have been possible, perhaps not conceivable. In others, technical developments were responses to the research questions.

Physical geography has experienced two parallel sets of methodological changes since 1970. The first involved closer alliances with other scientific disciplines, engaging with the physical, chemical, and biological bases for understanding physical matter and processes together with the mathematical methods necessary for their analysis. The second involved technical developments in field and laboratory measurement and data analysis. These two have come to pervade all work in physical geography, which has become technically sophisticated and whose progress has depended almost entirely on such skills.

Virtually all work in physical geography shares a belief in what is known as the "real" world—that which can be observed, measured, and generalized upon, even if the appreciation of particular events and landforms requires setting general principles within particular contexts. The laws of physics can be used to generalize about atmospheric processes, for example, but only an appreciation of how they interact in specific, local circumstances can account for the weather at a place on a given day. Immanent laws operate in local, contingent circumstances, involving highly complex interactions whose analysis requires sophisticated mathematical skills in analyzing nonlinear, often chaotic, relationships.

A much wider range of approaches is deployed within human geography; different theories of knowledge and reality inspire different types of work. The tenets of positivism still underpin some work in many areas: there is order in the world that can be observed, measured, analyzed, and generalized, even if there are no general laws of human behaviour awaiting discovery. Other work is based on theories of knowledge that claim an inseparability of observer and observed (or subject and object) and dispute the existence of real worlds independent of their inhabitants' imagined worlds. We cannot apprehend an external world but only perceived worlds. Geographical research based on these premises deploys means of identifying those worlds, the processes involved in their creation, and the behaviour within them. It then has to transmit that derived understanding to others—what is sometimes termed a "double hermeneutic".

These various approaches pervade most of contemporary human geography. With the exception of cultural geography, quantitative methods are used to analyze and identify regularities in data sets large and small, taking advantage of technical advances, such as with methods of artificial intelligence for classifying individuals and areas.

Nonquantitative approaches can be found throughout the various subdisciplines. These involve obtaining information in rigorous ways from individuals regarding their mental maps of the world and how these underpin behaviour. Means of interviewing individuals and groups to elicit information dominate the qualitative procedures that involve interpersonal interaction. Research material is also sought in a variety of other ways, through, for example, participant observation in case studies of communities and events. But information gathering extends well beyond interacting, directly or indirectly, with living people. Learning about the roles of places, spaces, and environments in the lives of individuals, groups, communities, and even entire societies near as well as far and past as well as present involves interrogating many information sources. Most common are written texts, analyzed for the meanings they can reveal. Other documents, such as maps, also reveal much, as do works of art. Ways of deconstructing meanings are commonly used in cultural and historical geography and in other subdisciplines too, as with the meanings attached to exotic foods in economic geography.

Research involves not only observing, recording, and analyzing the world but also transmitting acquired understandings and explanations to others. In quantitative analyses, this involves using mathematical notation and procedures—a language that many claim is unambiguous but whose use nearly always involves interpretation in vernacular languages, with meanings often contested. In qualitative work, nearly all of the reporting is done through the medium of written language. Having studied texts to reach understandings, researchers then deploy the same media to present them to others and thereby place their readers in the same situation of having to derive meanings from the textual material. The research process thus involves continued interpretation and reinterpretation of textual and other materials, including research reports. Unlike the apparently incontestable clear statements of quantitatively expressed research findings, research in much contemporary human geography involves continued debate over meanings and interpretations.

One tool long considered central to geographical work is the map. Automation of map production has been accompanied by a decline of research in this area; one of the few continuing fields concerns map legibility—the degree to which different symbols and shading succeed in transmitting messages. Its replacement as a central tool is GIS, a visualization medium with massive capacity for facilitating a wide range of research investigations. It offers not only sophisticated procedures for manipulating spatial data but also new ways of presenting visual data, including three-dimensional images of the world, at all scales. Geographic information science incorporates the traditional disciplines of cartography, geodesy, and photogrammetry with modern developments in remote sensing, the Global Positioning System (GPS), geostatistics, and geocomputation in activities that bring forward geographers' eternal interest in maps as sophisticated means of representing, analyzing, and viewing the Earth's great diversity.

Applied Geography

One area that some have set apart from the various subdisciplinary divisions concerns the application of geographical scholarship. Geography was always applied, long before it became an identified academic discipline; much geographical knowledge was created for specific purposes. Since the discipline was established, individuals have used their knowledge in a wide range of contexts and for various types of clients. Outside of universities, some of those trained as geographers have applied their skills in a range of sectors; the U.S. State Department had an Office of the Geographer for much of the 20th century, for example, providing the president with daily briefings.

For the first half of the 20th century, the development of geography as an academic discipline was closely associated with its educational role, especially in the preparation of teachers and of teaching materials. Increasingly, however, geographers responded to societal changes—especially the extending role of the state—by promoting their discipline as a potential contributor in a range of activities. Some, like L. Dudley Stamp, argued that geographers' factual knowledge regarding environments and places plus their understanding of spatial organization principles should be applied in town, city, country, and regional planning. This could just involve information provision, but increasingly it was argued that geographical analyses could inform the understanding of current patterns and trends and the preparation of plans for the future.

Such geographical involvement expanded in the late 20th century as pressures grew on universities to orient their work more to societal needs and to undertake applied research for public- and private-sector sponsors. Within human geography, for example, the locational analysis paradigm was adapted to commercial applications. Models of least-cost (and hence economically most efficient) location were used to predict the best sites for facilities, such as supermarkets and hospitals. Classifications of residential areas within cities were adapted to identify districts dominated by people with particular lifestyles toward which niche-market advertising could be directed; this substantial activity is widely termed geodemographics. Qualitative research findings and methods have been deployed in resolving conflicts over proposed land uses at particular sites.

Physical geographers' understanding of environmental processes has been directed to applied ends to meet concerns over environmental issues; much public policy takes these issues into account when pursuing goals such as sustainable development. Four types of applied work have been identified: description and auditing of contemporary environmental conditions; identification and analysis of environmental impacts, mainly of human action, actual and proposed; evaluation of the value of particular environments for specified future uses; and prediction and design of environmental works.

Some of these studies are relatively small-scale, such as tracing the diffusion of pollutants through water channels, identifying mineral deposits within local ecosystems, and monitoring local environmental changes and processes. Others involve larger-scale activities, such as models of climate change used to predict future ice-sheet melting, sea levels, and limits of cultivation of various plants. The scientific research may feed wider debates over policy formulation or may incorporate action plans—for conserving specific landscapes (such as wetlands or coasts) or managing a river catchment—as shown through the work of physical geographer William L. Graf, who chaired such interdisciplinary National Research Council studies.

The Geography of Contemporary Geography

The study of geography has changed considerably since its 19th-century institutionalization as an academic discipline, but several basic metaphors have been constant foundations of its endeavours. The first is of the world as a mosaic of patterns and forms, a complex map of myriad small areas with particular characteristics reflecting the interaction of environmental conditions and human activities. Much geographical scholarship has involved mapping that mosaic in all its variety and detail and conveying the observed areal differentiation of the Earth's surface to a wide audience. A second metaphor is of the world as a machine, comprising a large number of complexly interacting systems in which everything is both cause and effect; identifying and representing those systems is the basis for understanding cause and effect in environmental and human systems.

A third metaphor presents the world as an organism, in which the whole is greater than the sum of the parts but which, in turn, comprises a large number of subsidiary organisms and local regions with similar characteristics. Researchers have identified these organic elements, places in which the concurrent presence of various phenomena creates something more than just the sum of their parts—hence the French notion of characteristic genres de vie for each pays. Associated with this is the world as a text metaphor, in which the landscape is among the texts interpreted to appreciate its creators' intentions and cultures. Finally, and linked to the previous two, there is the metaphor of the world as an arena, with places as the contexts within which events occur: places are the contexts for learning and behaviour.

These metaphors are not mutually exclusive, and combinations of one or more are common. They are the contexts—or worldviews—within which scholarship is undertaken. Their relative importance varies over time and space; geography is a wide range of related academic practices reflecting local conditions in which geographers (individually and collectively) respond to their contexts. There may be common features—concerns reflecting the key concepts of environment, space and place, for example, and concentration on particular metaphors—but also local emphases and absences. In pre-Soviet Russia, for example, physical geographers stressed climatic variations and their influences on soils rather than on landforms as was typical elsewhere, and during the Soviet era human geography was largely absent, with just a few economic concerns of relevance to national planning having been studied.

Much international variation in geographical practices is set within the map of separate language realms. Each major national school has influenced the practice of geography in a number of others, some through their imperial projects. German and French influences have been strong in different parts of the Iberian world: in Latin America, German geographers influenced early development in Argentina, with a Catalan geographer having considerable influence in Venezuela and a Spaniard inaugurating developments in Panama. Japanese geography initially reflected German influences, in part refracted through American interpretations, especially at Berkeley; after 1945, physical and human geography were almost completely separated in Japan, with American influence dominating the latter. There has been growing concern internationally regarding the dominant role of English—and hence geographers in Anglophone countries—in the discipline's discourse.

Even within individual language realms, however, significant differences between the United Kingdom and the United States reflect important local contexts, despite many commonalities reflecting the substantial interchange across the Atlantic during the last half century. A major basis of those differences is geography's role in their educational systems.

The paucity of geographic education in schools in the United States was highlighted in the second half of the 20th century by the geographical ignorance of many Americans. Changing this situation was a cause taken up by several bodies. In the 1960s and '70s the National Science Foundation funded programs to upgrade science teaching, which included the American Association of Geographers' High School Geography Project. In the last decade of the 20th century, the National Geographic Society (internationally known for its National Geographic Magazine) committed substantial resources to promote geography in the country's schools, as well as launching a television channel to carry educational materials about human-environment interactions.

These major differences between the two countries are reflected in the pattern of specialisms within geography departments. In the United States, for example, there has been an increasing awareness that students can be attracted to undergraduate geography courses that provide training in marketable skills. Many departments have identified GIS as an important skills package, and increasing numbers of faculty appointments are of GIS specialists. In the United Kingdom such pressures are less, and cultural geography is more important; indeed, it dominates human geography in some departments, with spatial analysis having only a minor place in the curriculum. Furthermore, because geography degree programs in Britain are built on much deeper foundations of geographical exposure, there is less pressure to cover a full range of subdisciplinary specialisms. In addition, given the importance of prescribed research excellence in the funding of universities there, the current tendency is to build up specialist research teams in certain areas only.

There is thus a geography of geography as an academic discipline, as these national particularities are reproduced many times over. There are also differences within countries. Few departments (even the largest in the United Kingdom) cover the full range of the current subdisciplines in their teaching programs, for example, let alone in their research concentrations. Most specialize, reflecting the interests of senior staff at particular times in their development and institutional decisions on resource allocation. Thus, the practice of geography as an academic discipline itself reflects its own fundamental precepts. There are general features that apply to most geography programs but also particularities that reflect local characteristics and individual decision making. In geography, as in so much else, place matters.

In many ways, geography as practiced today is unrecognizable from the academic discipline that was being created at the end of the 19th century. And yet the underlying basic concepts—of environments, spaces, and places—remain at the disciplinary core. Geography continues to illuminate major aspects of the human condition through people's interactions with their natural and social milieux.

IMPORTANCE OF GEOGRAPHY

Even though educating our children about geography should remain important, it has slowly disappeared from the focus of teaching. Students, however, still need to understand how and where they fit into our nation and within the world.

Create Awareness of Place

Studying geography creates an awareness of place. Just like our founding fathers identified, understanding geography instills an identity of the American place.

More importantly, understanding geography helps us make sense of current and historical events, whether of economic, political, or social importance. We become better critical thinkers knowing this information. Geography pervades just about every aspect of our lives.

Develop Non-fiction Reading Skills

Studying geography develops non-fiction reading skills. Geography uses complex visual representations such as maps, pictures, charts, and graphs that must be interpreted depending on the purpose. Students of geography must use higher order thinking skills to analyze and synthesize information. Studying geography also naturally develops a working knowledge of how to read and process non-fiction text features since those features are woven throughout all aspects of the content. Finally, studying geography builds important vocabulary and background knowledge about our country and world too. If you want to build non-fiction reading skills, geography is a great resource.

Develop Spatial Awareness

Studying geography develops better spatial awareness. It is important to learn map sense and globe-reading skills yet these activities are virtually gone from education today. Could your students point north if you asked them? Could they look at a map and identify the location of our continents and oceans without the help of technology? Would your students know how to find their way home or to the next state without a GPS? By studying geography and mapping skills, we foster the development of spatial awareness and also create the link to understanding the effectiveness of key spatial geographic systems such as GPS. How will we improve on these technologies without another generation of students who understand how our world is structured?

Create a Global Community

Studying geography creates a global citizen. Those who know geography better understand the interdependence of our world and how we are connected through location, place, movement, region, and human-environment interactions. Think about it. As we develop our understanding of the important themes of geography, we also help build awareness for cultural diversity–how and why people live the way they do. Students need to understand this information in our global society.

Geography was and is still very important. And even though it may not be a part of your testing cycle, there are so many valuable reasons why integrating geography into your classroom is important.

NATURE OF GEOGRAPHY

Geography is concerned with place. Understanding the nature and causes of aerial differentiation on the global surface has been the geographer's task since people first noticed differences between places.

Through geography we seek to understand these differences in patterns of human distribution, interrelationships between human society and the physical environment, people's use of the Earth in time and space, and how these differences are related to people's cultures and economies. These, and other related themes, express major concerns of our time and reflect the consequences of spatial decisions.

In geography's pursuit of this understanding the questions "where?," "why?," and "how?" are central. The first of these introduces the issues of location and spatial choice; the latter two signify that modern geography is not content merely to describe, but seeks to explain. Beyond these questions, geographers also ask a fourth–"what if?"–as a means of seeking alternatives and giving the subject an applied dimension that can assist decision makers in planning and development at a variety of geographical scales.

The idea of place is not an examinable objective but an ultimate goal, whose pursuit gives direction to geographical study. As a geographical concept it refers to the aerial context of events, objects, and actions; in other words, to the patterns resulting from human occupancy of the global surface over time. The areal context is set in space which, though measurable, has by itself no meaning. Space becomes place when humans invest it with meaning, most commonly by giving it a name and all of the associations that that name evokes. The terms place and region may, therefore, be distinguished by spatial scale, rather than by inherent differences, for both involve space that has been invested with meaning. Two important implications flow from this understanding of place:

First, geography is strongly influenced by the norms of the social sciences. The complexity and changing nature of human society seldom permit the type of precision expected in the physical sciences. Instead, the social sciences offer a variety of perspectives and methods of study by which to examine the consequences of human behaviour on the global surface. In studying the idea of place from a spatial perspective, geographers inevitably encounter the problem of change through time; for them, landscape is document. Historians, too, are concerned with change through time as they document the consequences of human behaviour. Thus, like geographers, historians are also concerned with place. Indeed, a common concern with place brings the work of geographers and historians close together.

Second, physical geographers, no less than human geographers, contribute to an understanding of place; for the concept of site—the physical characteristics of a place—is integral to understanding aerial differentiation on the global surface. Nevertheless, geographers focus on the patterns and interactions to be found on that surface, and not primarily on the natural processes that act on it from above or below. They recognize that interaction between humans and their environment has always been mutual, and that the growth of technology has increased the human capacity to modify the environment.

That growth in technology has greatly aided geographers in their traditional tasks. It has given them increasingly refined techniques for gathering and interpreting data, whether in the field by means of GPS or by aerial and satellite imagery. Spatial relationships are at the heart of geography. Using software to analyze spatial relationships among objects being mapped, GIS, in particular, has greatly assisted geographers in depicting the character of place. Not only can they now process larger quantities of data more quickly and with greater refinement, but also they can manipulate variables and thus project alternatives that give geography an applied dimension. Finally, their work can be displayed using advanced techniques of computer-generated mapping. The view of geography presented here is that of a core sharply focused on the concept of place one in which both physical and human elements play an important part. The subject has an applied dimension that can affect our daily lives. It can, therefore, be a powerful medium for the development of skills contributing to citizenship and cultural awareness.

2

Human Geography

Human geography is concerned with the study of people and their communities, cultures, economies, and interactions with the environment. It includes economic geography, historical geography, political geography, social geography, urban geography, etc. All these divisions under human geography have been carefully analyzed in this chapter.

Human geography is one of the two major branches of geography, together with physical geography. Human geography is also called cultural geography. It is the study of the many cultural aspects found throughout the world and how they relate to the spaces and places where they originate and the spaces and places they then travel to, as people continually move across various areas.

Some of the main cultural phenomena studied in human geography include language, religion, different economic and governmental structures, art, music, and other cultural aspects that explain how and why people function as they do in the areas in which they live. Globalization is also becoming increasingly important to the field of human geography as it is allowing these specific aspects of culture to travel across the globe easily.

Cultural landscapes are important to the field because they link culture to the physical environments in which people live. A cultural landscape can either limit or nurture the development of various aspects of culture. For instance, people living in a rural area are often more culturally tied to the natural environment around them than those living in a large metropolitan area. This is generally the focus of the "Man-Land Tradition" in the Four Traditions of geography, which studies the human impact on nature, the impact of nature on humans, and people's perception of the environment.

Human geography developed out of the University of California, Berkeley and was led by Professor Carl Sauer. He used landscapes as the defining unit of geographic study and said that cultures develop because of the landscape and also, conversely, help to develop the landscape. Sauer's work and the cultural geography of today are highly qualitative in contrast to the quantitative methodology used in physical geography.

Human geography is still practiced, and more specialized fields within it have developed to further aid in the study of cultural practices and human activities as they relate spatially to the world. Such specialized fields include feminist geography, children's geography, tourism studies, urban geography, the geography of sexuality and space, and political geography.

Feminist Geography

Feminist geography is a sub-discipline of human geography that applies the theories, methods, and critiques of feminism to the study of the human environment, society, and geographical space. Feminist geography emerged in the 1970s, when members of the women's movement called on academia to include women as both producers and subjects of academic work. Feminist geographers aim to incorporate positions of race, class, ability, and sexuality into the study of geography. The discipline has been subject to several controversies.

The Geography of Women

The geography of women focuses on the describing the effects geography has on gender inequality and is theoretically influenced by welfare geography and liberal feminism. Feminist geographers emphasize the various gendered constraints put in place by distance and spatial separation (for instance, spatial considerations can play a role in confining women to certain locations or social spheres). In their book Companion to Feminist Geography, Seager and Johnson argue that gender is only a narrow-minded approach to understanding the oppression of women throughout the decades of colonial history. As such, understanding the geography of women requires a critical approach to questions of the dimensions of age, class, ethnicity, orientation and other socio-economic factors. An early objection to the concept of geography of women, however, claimed that gender roles were mainly explained through gender inequality. However, Foord and Gregson argue that the idea of gender roles emerges from a static social theory that narrows the focus to women and portrays women as victims, which gives a narrow reading of distance. Instead, they claim that the concept of the geography of women is able to display how spatial constraint and separation enter into the construction of women's positions. In 2004, theorist Edward Said critiqued the idea of geographical spaces in such a context where actions on gendered practices of representation are fabricated through dominant ideological beliefs. In response, feminist geographers argue that misrepresentations of gender roles and taken-for-granted feminist movements reveal that the challenges of the colonial present lie within the confinement of women to limited spatial opportunities. Therefore, feminist geographies are built on the principle that gender should be applied and developed in terms of space.

Socialist Feminist Geography

Socialist feminist geography, theoretically influenced by Marxism and Socialist feminism, seeks to explain inequality, the relationship between capitalism and patriarchy, and the interdependence of geography, gender relations, and economic development under capitalism. Socialist feminist geography revolves around questions of how to reduce the gender inequality caused by patriarchy and capitalism, and focuses predominantly on spatial separation, gender place, and locality. Uncertainty regarding appropriate articulation of gender and class analysis fuels a key theoretical debate within the field of socialist feminist geography. For example, when analyzing married mainland Chinese immigrant women living in New York City, women remain the primary object of analysis, and gender remains the primary social relation. However, socialist feminist geographers also recognize that many other factors, such as class, affect women's post-migration experiences and circumstances.

Socialist feminist geographers first worked primarily at the urban scale: Anglo-American feminist geographers focused on the social and spatial separation of suburban homes from paid employment. This was seen as vital to the day-to-day and generational development and maintenance of traditional gender relations in capitalist societies.

Socialist feminist geographers also analyze the ways in which the effects of geographical differences on gender relations not only reflect, but also partly determine local economic changes. Judith Butler's concept of "citationality" explores the lack of agency surrounding the facilitation of the presence of women within the discipline of geography. Subsequently, feminist geographers conclude that whenever performative measures are taken to diminish women's rights in geographical space, surrounding conventions adapt to make it seem as the norm.

Feminist Geographies of Difference

Feminist geographies of difference is an approach to feminist geography that concentrates on the construction of gendered identities and differences among women. It examines gender and constructions of nature through cultural, post-structural, postcolonial and psychoanalytic theories, as well as writings by women of color, lesbian women, gay men, and women from third world countries. In this approach, feminist geographers emphasize the study of micro-geographies of body, mobile identities, distance, separation and place, imagined geographies, colonialism and post-colonialism, and environment or nature.

Since the late 1980s, many feminist geographers have moved on to three new research areas: categories of gender between men and women, the formation of gender relations and identities, and the differences between relativism and situational knowledge.

Firstly, feminist geographers have contested and expanded the categories of genders between men and women. Through this, they have also begun to investigate differences in the constructions of gender relations across race, ethnicity, age, religion, sexuality and nationality, paying particular attention to women who are positioned along multiple axes of difference.

Secondly, to gain a better understanding of how gender relations and identities are formed and assumed, feminist geographers have drawn upon a broader extent of social theory and culture. Building upon this theoretical platform, feminist geographers are more able to discuss and debate the influence that post-structuralist and psychoanalytic theories have on multiple identifications.

Lastly, the difference between relativism and situated knowledge is a key area of discussion. Through these discussions, feminist geographers have discovered ways to reconcile partial perspectives with a commitment to political action and social change.

Critical Human Geography

Critical human geography is defined as "a diverse and rapidly changing set of ideas and practices within human geography linked by a shared commitment to emancipatory politics within and beyond the discipline, to the promotion of progressive social change and to the development of a broad range of critical theories and their application in geographical research and political practice".

Critical human geography emerged from the field of Anglophonic geography in the mid-1990s, and it presents a broad alliance of progressive approaches to the discipline. Critical human geographers focus on key publications that mark different eras of critical human geography, drawing upon anarchism, anti-colonialism, critical race theory, environmentalism, feminism, Marxism, nonrepresentational theory, post-Marxism, post-colonialism, post-structuralism, psychoanalysis, queer theory, situationism, and socialism.

Critical human geography is understood as being multiple, dynamic, and contested.

Rather than a specific sub-discipline of geography, feminist geography is often considered part of a broader, postmodern, critical theory approach, that draws upon the theories of Michel Foucault, Jacques Derrida, and Judith Butler, and many post-colonial theorists. Feminist geographers often focus on the lived experiences of individuals and groups in the geographies of their own localities, rather than theoretical development without empirical work.

Many feminist geographers study the same subjects as other geographers, but focus specifically on gender divisions. This has developed into concerns with wider issues of gender, family, sexuality, race, and class. Examples of areas of focus include:

- Geographic differences in gender relations and gender equality.
- The geography of women (e.g. spatial constraints and welfare geography).
- The construction of gender identity through the use and nature of spaces and places.
- Geographies of sexuality (queer theory).
- Children's geographies.

Feminist geographers are also deeply impacted by and respondent to contemporary globalization and neoliberal discourses that are manifested transnationally and translocally.

Feminist geography also critiques human geography and other academic disciplines, arguing that academic structures have been traditionally characterized by a patriarchal perspective and that contemporary studies which do not confront the nature of previous work reinforce the masculine bias of academic study. British geographer Gillian Rose's Feminism and Geography is one such sustained criticism that claims that the approach to human geography in Britain is historically masculinist. This geographic masculinization includes traditions of writing landscapes as feminine spaces—and thus as subordinate to male geographers—and subsequent assumptions of a separation between mind and body. Johnston & Sidaway describe such separation as "Cartesian dualism" and further explain its influence on geography:

> 'Cartesian dualism underlines our thinking in a myriad of ways, not least in the divergence of the social sciences from the natural sciences, and in a geography which is based on the separation of people from their environments. Thus while geography is unusual in its spanning of the natural and social sciences and in focusing on the interrelations between people and their environments, it is still assumed that the two are distinct and one acts on the other. Geography, like all of the social sciences, has been built upon a particular conception of mind and body which sees them as separate, apart and acting on each other '

Thus, too, feminist work has sought to transform approaches to the study of landscape by relating it to the way that it is represented ('appreciated' so to speak), in ways that are analogous to the heterosexual male gaze directed towards the female body. Both of these concerns (and others)- about the body as a contested site and for the Cartesian distinction between mind and body - have been challenged in postmodern and post-structuralist feminist geographies.

Other feminist geographers have interrogated how the discipline of geography itself represents and reproduces the heterosexual male gaze. Feminist geographers such as Katherine McKittrick have asserted that how we see and understand space are fundamentally bound up in how we understand the hegemonic presence of the white male subject in history, geography and in the materiality of everyday space. Building off of Sylvia Wynter's theories of the racialized production of public and private space, McKittrick challenges "social landscapes that presume subaltern populations have no relationship to the production of space" and writes to document black female geographies in order to "allow us to engage with a narrative that locates and draws on black histories and black subjects in order to make visible social lives which are often displaced, rendered ungeographic". McKittrick stakes claim in the co-articulation of race and gender as they articulate space, writing: "I am emphasizing here that racism and sexism are not simply bodily or identity-based; racism and sexism are also spatial acts and illustrate black women's geographic experiences and knowledges as they are made possible through domination". Moreover, many feminist geographers have critiqued human geography for centering masculine knowledge emphasizing "objective" knowledge, arguing instead for the use of situated knowledge which understands both observation and analysis as being rooted in partial objectivity.

Challenges of Feminist Geography

Linda McDowell and Joanne P. Sharp, both foundational feminist geographers and scholars, describe the struggle of gaining recognition in academia, saying that "it has been a long struggle to gain recognition within geography as a discipline that gender relations are a central organizing feature both of the material and symbolic worlds and of the theoretical basis of the discipline'.' Feminist geographers struggle in academia in a variety of ways. Firstly, ideas that originate from feminist discourse are often seen as commonsense once the wider field accepts them, thereby rendering geography that is explicitly feminist invisible. Furthermore, feminist geography is understood to be the only subfield of geography where gender is explicitly addressed, permitting the wider discipline to disengage from feminist challenges. Finally, within the field, some geographers believe that feminist practice has been fully integrated into the academy, making feminist geography obsolete.

Challenges of feminist geography are also embedded in the subfield itself. The epistemology of feminist geography argues that the positionalities and lived experiences of the geographers are as central to scholarship as what is being researched. In this way, feminist geographers must maintain diverse identities to fully engage with the discipline. Linda Peake and Gill Valentine point out that, while feminist geography has addressed gender issues in more than twenty-five countries across the world, scholarship in the field of feminist geography is primarily conducted by white female scholars from institutions in the Global North. In this way, feminist geography faces not only barriers rooted in the academy but a lack of diversity in its own field.

Feminist geographers draw upon a broad range of social and cultural theory, including psychoanalysis and post-structuralism, to develop a fuller understanding of how gender relations and identities are shaped and assumed. This has led to the fundamental rethinking of gender and the contradictions and possibilities presented by the seeming instability and insistent repetitions of gender norms in practice. The focus on multiple identifications and the influence of post-structuralist and psychoanalytic theories has allowed feminist geographers to enter into dialogue with other strands of critical geography. This open dialogue, however, has also allowed tensions to build between geographers in the United States and geographers in Great Britain. Theoretical differences among feminist geographers are more obvious than in the past, but since 1994, the national differences between America and British geographers have begun to diminish as both parties pursue new directions.

Controversies Surrounding Feminist Geography

In 2018, a leading journal in feminist geography entitled Gender, Place and Culture, was subject to a scholarly publishing hoax. Helen Pluckrose, James A. Lindsay, and Peter Boghossian submitted a paper titled "Human Reactions to Rape Culture and Queer Performativity in Urban Dog Parks in Portland, Oregon". The paper proposed that dog parks are "rape-condoning spaces", and a place of rampant canine rape culture and systemic oppression against "the oppressed dog" through which human attitudes to both problems can be measured and analysed by applying black feminist criminology. The paper suggested that this could provide insight into training men out of the sexual violence and bigotry. The paper has since been retracted. The hoax has been criticized as unethical and mean-spirited, and critics of the hoax have suggested that the hoaxers did not understand the process of peer review.

POPULATION GEOGRAPHY

Population geography is a division of human geography. It is the study of the ways in which spatial variations in the distribution, composition, migration, and growth of populations are related to the nature of places. Population geography involves demography in a geographical perspective. It focuses on the characteristics of population distributions that change in a spatial context. This often involves factors such as where populations are found and how the size and composition of these populations is regulated by the demographic processes of fertility, mortality, and migration. Contributions to population geography are cross-disciplinary because geographical epistemologies related to environment, place and space have been developed at various times. Related disciplines include geography, demography, sociology, and economics.

Since its inception, population geography has taken at least three distinct but related forms, the most recent of which appears increasingly integrated with human geography in general. The earliest and most enduring form of population geography emerged in the 1950s, as part of spatial science. Pioneered by Glenn Trewartha, Wilbur Zelinsky, William A. V. Clark, and others in the United States, as well as Jacqueline Beujeau-Garnier and Pierre George in France, it focused on the systematic study of the distribution of population as a whole and the spatial variation in population characteristics such as fertility and mortality. Population geography defined itself as the systematic study of:

- The simple description of the location of population numbers and characteristics.

- The explanation of the spatial configuration of these numbers and characteristics.

- The geographic analysis of population phenomena (the inter-relations among real differences in population with those in all or certain other elements within the geographic study area).

Accordingly, it categorized populations as groups synonymous with political jurisdictions representing gender, religion, age, disability, generation, sexuality, and race, variables which go beyond the vital statistics of births, deaths, and marriages. Given the rapidly growing global population as well as the baby boom in affluent countries such as the United States, these geographers studied the relation between demographic growth, displacement, and access to resources at an international scale.

Examples can be shown through population density maps. A few types of maps that show the spatial layout of population are choropleth, isoline, and dot maps:

- Demographic phenomena (natality, mortality, growth rates, etc.) through both space and time.

- Increases or decreases in population numbers.

- The movements and mobility of populations.

- Occupational structure.

- The way in which places in turn react to population phenomena, e.g. immigration.

Research topics of other geographic sub-disciplines, such as settlement geography, also have a population geography dimension:

- The grouping of people within settlements.

- The way from the geographical of places, e.g. settlement patterns.

All of the above are looked at over space and time. Population geography also studies human-environment interactions, including problems from those relationships, such as overpopulation, pollution, and others.

Historical Geography

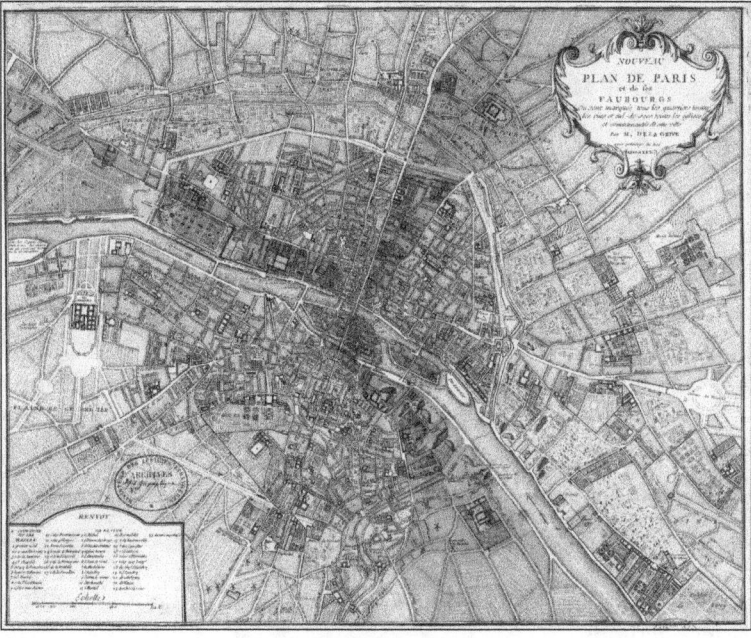

A 1740 map of Paris.

Historical geography is the branch of geography that studies the ways in which geographic phenomena have changed over time. It is a synthesizing discipline which shares both topical and methodological similarities with history, anthropology, ecology, geology, environmental studies, literary studies, and other fields. Although the majority of work in historical geography is considered human geography, the field also encompasses studies of geographic change which are not primarily anthropogenic. Historical geography is often a major component of school and university curricula in geography and social studies. Current research in historical geography is being performed by scholars in more than forty countries.

Historical geography seeks to determine how cultural features of various societies across the planet emerged and evolved by understanding their interaction with their local environment and surroundings.

Development of the Discipline

In its early days, historical geography was difficult to define as a subject. A textbook from the 1950s cites a previous definition as an 'unsound attempt by geographers to explain history'. Its author, J. B. Mitchell, came down firmly on the side of geography: 'the historical geographer is a geographer first last and all the time'. By 1975 the first number of the Journal of Historical Geography had widened the discipline to a broader church: 'the writings of scholars of any disciplinary provenance who have something to say about matters of geographical interest relating to past time'.

For some in the United States of America, the term historical geography has a more specialized meaning: the name given by Carl Ortwin Sauer of the University of California, Berkeley to his program of reorganizing cultural geography (some say all geography) along regional lines, beginning in the first decades of the 20th century. To Sauer, a landscape and the cultures in it could only be understood if all of its influences through history were taken into account: physical, cultural, economic, political, environmental. Sauer stressed regional specialization as the only means of gaining sufficient expertise on regions of the world. Sauer's philosophy was the principal shaper of American geographic thought in the mid-20th century. Regional specialists remain in academic geography departments to this day. Despite this, some geographers feel that it harmed the discipline; that too much effort was spent on data collection and classification, and too little on analysis and explanation. Studies became more and more area-specific as later geographers struggled to find places to make names for themselves. These factors may have led in turn to the 1950s crisis in geography, which raised serious questions about geography as an academic discipline in the USA.

This sub-branch of human geography is closely related to history, environmental history, and historical ecology.

Demography

Demography is a statistical study of human populations , especially with reference to size and density, distribution, and vital statistics (births, marriages, deaths, etc.). Contemporary demographic concerns include the "population explosion," the interplay between population and economic development, the effects of birth control, urban congestion, illegal immigration, and labour force statistics.

The roots of statistical demography may be found in the work of the Englishman John Graunt; his work examines the weekly records of deaths and baptisms (the "bills of mortality") dating back to the end of the 16th century. In search of statistical regularities, Graunt made an estimate of the male-female ratios at birth and death-birth ratios in London and rural communities. His most celebrated contribution was his construction of the first mortality table; by analyzing birth and death rates he was able to estimate roughly the number of men currently of military age, the number of women of childbearing age, the total number of families, and even the population of London. Another such study was undertaken by Johann Süssmilch, analyzed the populations of 1,056 parishes in Brandenburg and various cities and provinces of Prussia. Süssmilch constructed several mortality tables, most notably the first such table for the whole population of Prussia.

In 18th-century Europe, the development of life insurance and growing attention to public health produced an increased awareness of the significance of mortality studies. Civil registries of significant public events (births, deaths, and marriages) began in the 19th century to supplant church registries. Censuses of the population also developed during the 19th century.

For most of the 19th century, demographic studies continued to emphasize the phenomenon of mortality; it was not until demographers noted that a considerable decline of fertility had taken place in the industrialized countries during the second half of the 19th century, that they began to study fertility and reproduction with as much interest as they studied mortality. The phenomenon of differential fertility, with its implications about selection and more particularly about the evolution of intelligence, evoked widespread interest as shown in Charles Darwin's theories and in the works of Francis Galton. During the period between the two world wars, demography took on a broader, interdisciplinary character. In 1928 the International Union for the Scientific Study of Population was founded.

In spite of increasing sophistication in the analysis of statistics and the proliferation of research institutes, periodicals, and international organizations devoted to the science of demographics, the basis for most demographic research continues to lie in population censuses and the registration of vital statistics. Even the most meticulously gathered census is not completely accurate, however, and birth, death, and marriage statistics—based on certificates drawn up by local authorities—are accurate mostly in countries with a long tradition of registry.

DEVELOPMENT GEOGRAPHY

Development geography is a branch of geography which refers to the standard of living and its quality of life of its human inhabitants. In this context, development is a process of change that affects people's lives. It may involve an improvement in the quality of life as perceived by the people undergoing change. However, development is not always a positive process. Gunder Frank commented on the global economic forces that lead to the development of underdevelopment. This is covered in his dependency theory.

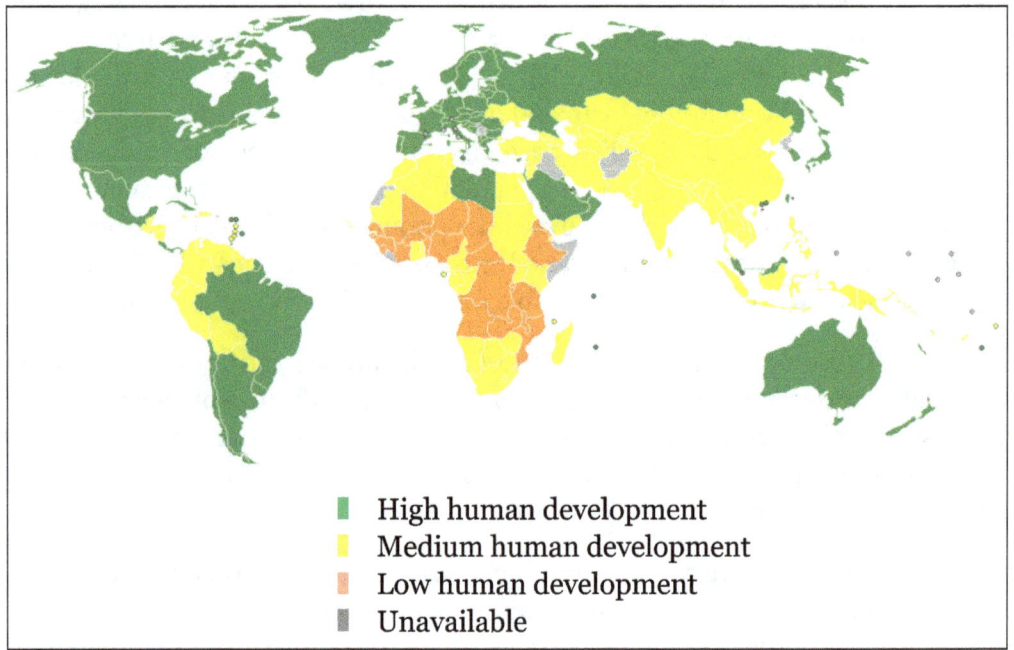

    ■ High human development
    ■ Medium human development
    ■ Low human development
    ■ Unavailable

In development geography, geographers study spatial patterns in development. They try to find by what characteristics they can measure development by looking at economic, political and social factors. They seek to understand both the geographical causes and consequences of varying development. Studies compare More Economically Developed Countries (MEDCs) with Less Economically Developed Countries (LEDCs). Additionally variations within countries are looked at such as the differences between northern and southern Italy, the Mezzogiorno.

Quantitative Indicators

Quantitative indicators are numerical indications of development:

- Economic include GNP (Gross National Product) per capita, unemployment rates, energy consumption and percentage of GNP in primary industries. Of these, GNP per capita is the most used as it measures the value of all the goods and services produced in a country, excluding those produced by foreign companies, hence measuring the economic and industrial development of the country. However, using GNP per capita also has many problems:

- It does not take into account the distribution of the money which can often be extremely unequal as in the UAE where oil money has been collected by a rich elite and has not flowed to the bulk of the country.

- GNP does not measure whether the money produced is actually improving people's lives and this is important because in many MEDCs there are large increases in wealth over time but only small increases in happiness.

- The figure rarely takes into account the unofficial economy, which includes subsistence agriculture and cash-in-hand or unpaid work, which is often substantial in LEDCs. In LEDCs it is often too expensive to accurately collect this data and some governments intentionally or unintentionally release inaccurate figures.

- The figure is usually given in US dollars which due to changing currency exchange rates can distort the money's true street value so it is often converted using purchasing power parity (PPP) in which the actual comparative purchasing power of the money in the country is calculated.

- Social indications include access to clean water and sanitation (which indicate the level of infrastructure developed in the country) and adult literacy rate, measuring the resources the government has to meet the needs of the people.

- Demographic indicators include the birth rate, death rate and fertility rate, which indicate the level of industrialization:

  - Health indicators (a sub-factor of demographic indicators) include nutrition (calories per day, calories from protein, percentage of population with malnutrition), infant mortality and population per doctor, which indicate the availability of healthcare and sanitation facilities in a country.

- Environmental indications include how much a country does for the environment.

Composite Indicators

- In the table below GDP stands for gross domestic product, which is generally taken to be equal to GNP.

- Other composite measures include the PQLI (Physical Quality of Life Index) which was a precursor to the HDI which used infant mortality rate instead of GNP per capita and rated countries from 0 to 100. It was calculated by assigning each country a score of 0 to 100 for each indicator compared with other countries in the world. The average of these three numbers makes the PQLI of a country.

- The HPI (Human Poverty Index) is used to calculate the percentage of people in a country who live in relative poverty. In order to better differentiate the number of people in abnormally poor living conditions the HPI-1 is used in developing countries, and the HPI-2 is used in developed countries. The HPI-1 is calculated based on the percentage of people not expected to survive to 40, the adult illiteracy rate, the percentage of people without access to safe water, health services and the percentage of children under 5 who are underweight. The HPI-2 is calculated based on the percentage of people who do not survive to 60, the adult functional illiteracy rate and the percentage of people living below 50% of median personal disposable income.

- The GDI (Gender-related Development Index) measures gender equality in a country in terms of life expectancy, literacy rates, school attendance and income.

HDI rank	Country	GDP per capita (PPP US$) 2008	Human development index (HDI) value 2006
4	Australia	35,677	0.965
70	Brazil	10,296	0.807
151	Zimbabwe	188	0.513

Qualitative Indicators

Qualitative indicators include descriptions of living conditions and people's quality of life. They are useful in analyzing features that are not easily calculated or measured in numbers such as freedom, corruption, or security, which are largely non-material benefits.

Geographic Variations in Development

There is a considerable spatial variation in development rates.

> Global wealth also increased in material terms, and during the period 1947 to 2000, average per capita incomes tripled as global GDP increased almost tenfold (from $US3 trillion to $US30 trillion). Over 25% of the 4.5 billion people in LEDCs still have life expectancies below 40 years. More than 80 countries have a lower annual per capita income in 2000 than they did in 1990. The average income in the world's five richest countries is 74 times the level in the world's poorest five, the widest it has ever been. Nearly 1.3 billion people have no access to clean water. About 840 million people are malnourished.

> — Stephen Codrington

The most famous pattern in development is the North-South divide. The North-South divide separates the rich North or the developed world, from the poor South. This line of division is not as straightforward as it sounds and splits the globe into two main parts. It is also known as the Brandt Line.

The "North" in this divide is regarded as being North America, Europe, Russia, South

Korea, Japan, Australia, New Zealand and the like. The countries within this area are generally the more economically developed. The "South" therefore encompasses the remainder of the Southern Hemisphere, mostly consisting of KFCs. Another possible dividing line is the Tropic of Cancer with the exceptions of Australia and New Zealand. It is critical to understand that the status of countries is far from static and the pattern is likely to become distorted with the fast development of certain southern countries, many of them NICs (Newly Industrialised Countries) including India, Thailand, Brazil, Malaysia, Mexico and others. These countries are experiencing sustained fast development on the back of growing manufacturing industries and exports.

Most countries are experiencing significant increases in wealth and standard of living. However, there are unfortunate exceptions to this rule. Noticeably some of the former Soviet Union countries has experienced major disruption of industry in the transition to a market economy. Many African nations have recently experienced reduced GNPs due to wars and the AIDS epidemic, including Angola, Congo, Sierra Leone and others. Arab oil producers rely very heavily on oil exports to support their GDPs so any reduction in oil's market price can lead to rapid decreases in GNP. Countries which rely on only a few exports for much of their income are very vulnerable to changes in the market value of those commodities and are often derogatively called banana republics. Many developing countries do rely on exports of a few primary goods for a large amount of their income (coffee and timber for example), and this can create havoc when the value of these commodities drops, leaving these countries with no way to pay off their debts.

Within countries the pattern is that wealth is more concentrated around urban areas than rural areas. Wealth also tends towards areas with natural resources or in areas that are involved in tertiary (service) industries and trade. This leads to a gathering of wealth around mines and monetary centres such as New York, London and Tokyo.

Barriers to International Development

Geographers along with other social scientists have recognized that certain factors present in a given society may impede the social and economic development of that society. Factors, which have been identified as obstructing the economic and social welfare of developing societies, include:

- Lack of education,

- Lack of healthcare,

- Pervasiveness of intoxicating drugs,

- Weak political, social, and economic institutions,

- Ineffective taxation,

- Environmental degradation,

- Lack of religious/gender/racial/sexual freedoms,

- Indebtedness,

- Protectionist barriers to trade,

- Foreign aid,

- Dependence upon primary resource exports,

- Unequal distribution of wealth,

- Inhospitable climate.

Effective governments may address many barriers to economic and social development, however in many instances this is challenging due to the path dependency societies develop regarding many of these issues. Some barriers to development may be impossible to address, such as climatic barriers to development. In these cases societies must evaluate whether such climatic barriers to development dictate that society must relocate a given settlement in order to enjoy greater economic development.

Many scholars agree that foreign aid provided to developing nations is ineffective and in many instances counter productive. This is due to the manner in which foreign aid changes the incentives for productivity in a given developing society, and the manner in which foreign aid has the tendency to corrupt the governments responsible for its allocation and distribution.

Cultural barriers to development such as discrimination based on gender, race, religion, or sexual orientation are challenging to address in certain oppressive societies, though recent progress has been significant in some societies.

While the aforementioned barriers to economic growth and development are most prevalent in the less developed economies of the world, even the most developed economies are plagued by select barriers to development such as drug prohibition and income inequality.

Aid

MEDCs (More Economically Developed Countries) can give aid to LEDCs (Less Economically Developed Countries). There are several types of aid:

- Governmental (bilateral) aid.

- International Organizational (multilateral) aid, e.g. The World Bank.

- Voluntary aid from individuals, often mediated through NGOs.

- Short-term/emergency aid (humanitarian assistance).

- Long-term/sustainable aid.

- Non-governmental organization (NGO) aid.

Aid can be given in several ways. Through money, materials, or skilled and learned people (e.g. teachers).

Aid has advantages. Mostly short-term or emergency aid help people in LEDCs to survive a natural (earthquake, tsunami, volcano eruption etc.) or human (civil war etc.) disaster. Aid helps make the recipient country (the country that receives aid) get more developed.

However, aid also has disadvantages. Often aid does not even reach the poorest people. Often money gained from aid is used up to make infrastructures (bridges, roads etc.), which only the rich can use. Also, the recipient country becomes more dependent on aid from a donor country (the country giving aid).

Whilst the above conception of aid has been the most pervasive within development geography work, it is important to remember that the aid landscape is far more complex than one directional flows from 'developed' to 'developing' countries. Development geographers have been at the forefront of research that aims to understand both the material exchanges and discourse surrounding 'South-South' development cooperation. 'Non-traditional' foreign aid from Southern, Middle Eastern and post-Socialist states (those outside the Development Assistance Committee (DAC) of the OECD) provide alternative development discourses and approaches to that of the mainstream Western model. Development geographers seek to examine the geopolitical drivers behind the aid donor programmes of "LEDCs", as well as the discursive symbolic repertoires of non-DAC donor states. Two illustrative examples of the complex aid landscape are that of China, which has been active as an aid donor throughout the latter half of the twentieth century but published its first report on foreign aid policy as recently as 2011 and India, an often cited aid recipient, but which has had donor programmes to Nepal and Bhutan since the 1950s.

SOCIAL GEOGRAPHY

Social geography is the branch of human geography that is most closely related to social theory in general and sociology in particular, dealing with the relation of social phenomena and its spatial components. Though the term itself has a tradition of more than 100 years, there is no consensus on its explicit content. In 1968, Anne Buttimer noted that "with some notable exceptions, social geography can be considered a field created and cultivated by a number of individual scholars rather than an academic tradition built up within particular schools". Since then, despite some calls for convergence centred on the structure and agency debate, its methodological, theoretical and topical diversity has spread even more, leading to numerous definitions of social geography and, therefore, contemporary scholars of the discipline identifying a great variety of different social geographies. However, as Benno Werlen remarked, these different perceptions are nothing else than different answers to the same two (sets of) questions, which refer to the spatial constitution of society on the one hand, and to the spatial expression of social processes on the other.

The different conceptions of social geography have also been overlapping with other sub-fields of geography and, to a lesser extent, sociology. When the term emerged within the Anglo-American tradition during the 1960s, it was basically applied as a synonym for the search for patterns in the distribution of social groups, thus being closely connected to urban geography and urban sociology. In the 1970s, the focus of debate within American human geography lay on political economic processes (though there also was a consider-able number of accounts for a phenomenological perspective on social geography), while in the 1990s, geographical thought was heavily influenced by the "cultural turn". Both times, as Neil Smith noted, these approaches "claimed authority over the 'social'". In the American tradition, the concept of cultural geography has a much more distinguished history than social geography, and encompasses research areas that would be conceptu-alized as "social" elsewhere. In contrast, within some continental European traditions, social geography was and still is considered an approach to human geography rather than a sub-discipline, or even as identical to human geography in general.

The term "social geography" (or rather "géographie sociale") originates from France, where it was used both by geographer Élisée Reclus and by sociologists of the Le Play School, perhaps independently from each other. In fact, the first proven occurrence of the term derives from a review of Reclus' Nouvelle géogra-phie universelle from 1884, written by Paul de Rousiers, a member of the Le Play School. Reclus himself used the expression in several letters, the first one dating from 1895, and in his last work L'Homme et la terre from 1905. The first person to employ the term as part of a publication's title was Edmond Demolins, anoth-er member of the Le Play School, whose article Géographie sociale de la France was published in 1896 and 1897. After the death of Reclus as well as the main proponents of Le Play's ideas, and with Émile Durkheim turning away from his early concept of social morphology, Paul Vidal de la Blache, who noted that ge-ography "is a science of places and not a science of men", remained the most in-fluential figure of French geography. One of his students, Camille Vallaux, wrote the two-volume book Géographie sociale, published in 1908 and 1911. Jean Brun-hes, one of Vidal's most influential disciples, included a level of (spatial) inter-actions among groups into his fourfold structure of human geography. Until the Second World War, no more theoretical framework for social geography was developed, though, leading to a concentration on rather descriptive rural and re-gional geography. However, Vidal's works were influential for the historical Annales School, who also shared the rural bias with the contemporary geographers, and Durkheim's concept of social morphology was later developed and set in connection with social geography by sociologists Marcel Mauss and Maurice Halbwachs.

The first person in the Anglo-American tradition to use the term "social geography" was George Wilson Hoke, whose paper The Study of Social Geography was published in 1907, yet there is no indication it had any academic impact. Le Play's work, however, was taken up in Britain by Patrick Geddes and Andrew John Herbertson. Percy M. Roxby, a former student of Herbertson, in 1930 identified social geography as one of human geography's four main branches. By contrast, the American academic geography of that time was dominated by the Berkeley School of Cultural Geography led by Carl O. Sauer, while the spatial distribution of social groups was already studied by the Chicago School of Sociology. Harlan H. Barrows, a geographer at the University of Chicago, nevertheless regarded social geography as one of the three major divisions of geography.

Another pre-war concept that combined elements of sociology and geography was the one established by Dutch sociologist Sebald Rudolf Steinmetz and his Amsterdam School of Sociography. However, it lacked a definitive subject, being a combination of geography and ethnography created as the more concrete counterpart to the rather theoretical sociology. In contrast, the Utrecht School of Social geography, which emerged in the early 1930s, sought to study the relationship between social groups and their living spaces.

In the German-language geography, this focus on the connection between social groups and the landscape was further developed by Hans Bobek and Wolfgang Hartke after the Second World War. For Bobek, groups of Lebensformen (patterns of life)—influenced by social factors—that formed the landscape, were at the center of his social geographical analysis. In a similar approach, Hartke considered the landscape a source for indices or traces of certain social groups' behaviour. The best-known example of this perspective was the concept of Sozialbrache (social-fallow), i.e. the abandoning of tillage as an indicator for occupational shifts away from agriculture.

Though the French Géographie Sociale had been a great influence especially on Hartke's ideas, no such distinct school of thought formed within the French human geography. Nonetheless, Albert Demangeon paved the way for a number of more systematic conceptualizations of the field with his (posthumously published) notion that social groups ought to be within the center of human geographical analysis. That task was carried out by Pierre George and Maximilien Sorre, among others. Then a Marxist, George's stance was dominated by a socio-economic rationale, but without the structuralist interpretations found in the works of some the French sociologists of the time. However, it was another French Marxist, the sociologist Henri Lefebvre, who introduced the concept of the (social) production of space. He had written on that and related topics since the 1930s, but fully expounded it in La Production de L'Espace as late as 1974. Sorre developed a schema of society related to the ecological idea of habitat, which was applied to an urban context by the sociologist Paul-Henry Chombart de Lauwe. For the Dutch geographer Christiaan van Paassen, the world consisted of socio-spatial entities of different scales formed by what he referred to as a "syn-ecological complex", an idea influenced by existentialism.

A more analytical ecological approach on human geography was the one developed by Edgar Kant in his native Estonia in the 1930s and later at Lund University, which he called "anthropo-ecology". His awareness of the temporal dimension of social life would lead to the formation of time geography through the works of Torsten Hägerstrand and Sven Godlund.

BEHAVIORAL GEOGRAPHY

Behavioral geography is an approach to human geography that examines human behavior using a disaggregate approach. Behavioral geographers focus on the cognitive processes underlying spatial reasoning, decision making, and behavior. In addition, behavioral geography is an ideology/approach in human geography that makes use of the methods and assumptions of behaviorism to determine the cognitive processes involved in an individual's perception of or response and reaction to their environment.

Behavioral geography is that branch of human science, which deals with the study of cognitive processes with its response to its environment, through behaviorism.

Issues

Because of the name it is often assumed to have its roots in behaviorism. While some behavioral geographers clearly have roots in behaviorism due to the emphasis on cognition, most can be seen as cognitively oriented. Indeed, it seems that behaviorism interest is more recent and growing. This is particularly true in the area of human landscaping.

Behavioral geography draws from early behaviorist works such as Tolman's concepts of "cognitive maps". More cognitively oriented, behavioral geographers focus on the cognitive processes underlying spatial reasoning, decision making, and behavior. More behaviorally oriented geographers are materialists and look at the role of basic learning processes and how they influence the landscape patterns or even group identity.

The cognitive processes include environmental perception and cognition, wayfinding, the construction of cognitive maps, place attachment, the development of attitudes about space and place, decisions and behavior based on imperfect knowledge of one's environs, and numerous other topics.

The approach adopted in behavioral geography is closely related to that of psychology, but draws on research findings from a multitude of other disciplines including economics, sociology, anthropology, transportation planning, and many others.

The Social Construction of Nature

Nature is the world which surrounds us, including all life (plants, animals, organisms, humans, etc.) and physical features. Social Construction is the way that human beings process the world around us in our minds. According to Plato's 'Classical Theory of Categorization', humans create categories of what they see through experience and imagination. Social constructionism, therefore, is this characterization that makes language and semantics possible. If these experiences and imageries are not placed into categories, then the human ability to think about it becomes limited.

The social construction of nature looks to question different truths and understandings for how people treat nature, based on when and where someone lives. In academic circles, researchers look at how truths exist (ontology) and how truths are justified (epistemology). Construction is both a process and an outcome, where people's understandings of the word nature can be both literal and metaphorical, such as through giving it a human quality (Mother Nature). It can also be used to discredit science or philosophy.

As a subset of behavioral geography, the social construction of nature also includes environmental ethics and values, which affect how humans treat, and interact with, the natural environment. It incorporates ideas from environmental science, ecology, sociology, geography, biology, theology, philosophy, psychology, politics, economics, and other disciplines, to bring together the social, cultural and environmental dimensions of life. Social constructionism uses a lot of ideas from Western world thinking, but it is also incorporates truths from other world views, such as the Traditional Knowledge of Aboriginal groups, or more specifically ecofeminism and cosmology in India or ubuntu philosophy in Africa, for example. It is also related to postmodernism and the concept of the Anthropocene, that views humans as a force that is redirecting the geological history of Earth, destroying nature.

The Role of Linguistics

There are many ways of understanding and interpreting nature. According to Raymond Williams, there are three ways to give meaning to (or define) nature:

- Nature as a quality, character or process (e.g. human nature).

- Nature as a force (e.g. weather).

- Nature as the material world (e.g. the physical environment).

According to Raymond Williams, language plays a role in how we understand, interpret, and give meaning to nature. This is how multiple truths can be valid at the same time.

The Role of Mental Maps

Humans have the ability to create images of their environments through experiences in their mind. These experiences allow us to create mental maps where we can create memories associated to space. It is a two-way process where the environment provides suggestions for what should be seen, and then the observer gives meaning with those suggestions.

These images have three parts:

- An identity.

- A pattern.

- A practical or emotional meaning.

According to Kevin Lynch, the environmental images (or mental maps) that we make can either be weak or strong, where the process is ongoing and never stops.

The Role of Science

Science occurs at many dimensions and scales that do not consider culture, but can be motivated by politics, economics and ethics. Scientific knowledge consists of concepts and analysis, and is a way to represent nature.

According to Michel Foucault, a truth does not have to be close to reality for it to be worth something or have power. For Carolyn Merchant, science can only be given power if a truth is interpreted as having worth.

Schools of Thought

Relativism is important in the social construction of nature, as all truths are relative to the perspective they are coming from. There are two schools of thought on how the social construction of nature is relative:

- Critical Realism (being realistic).

- Pragmatism (being practical).

Critical realists reject the idea of relativism and rely more on natural sciences. Pragmatists have no set opinion on the matter and rely on social science and ethics, instead.

According to Richard Rorty, relativism is relevant to pragmatism in three ways:

- Every belief is equally valid.

- There is no criteria for what a truth can be.

- That any truth can be justified by the society it comes from.

According to Gilbert White, pragmatism has four main assumptions:

- That human existence is based on putting labor into the land.

- That the idea of owning anything is a conception.

- That humans learn from their experiences.

- That engagement of the publics is what allows for commitments.

Richard Rorty also associated three characteristics to pragmatism:

- That all theories characterize some form of truth.

- That there is not difference between what can and should be done when it comes to the truth.

- That knowledge is constrained by the conversations we have.

Being pragmatic is the more accepted school of thought for social construction being a relative concept.

How Nature becomes Socially Constructed

Nature can be socially constructed by both culturally interpreting and physically shaping the environment. This can happen in three ways:

- Using non-human symbols to represent nature (Totemism).

- Using non-human animals to relate to nature (Animism).

- Viewing nature as an 'Other' (Naturalism).

Constructions can also be categorized by giving them meaning through the process of embodiment, which has three components:

- The 'habitus' (the individual).

- The practice it originates from (the culture).

- An associated taxonomic group (i.e. homo sapiens).

No matter how nature becomes socially constructed, though, the process itself is limited by three dimensions:

- The physical dimension.

- The mental dimension.

- The social dimension.

The physical dimension is limited to the human body, where the brain is responsible for creating and selecting thoughts. The mental dimension is used to understand the physical dimension and is limited to human logic. The social dimension needs moral and social order and is used to give meaning to both what is physically present and what is culturally constructed. All three dimensions must be present and linked to be able to socially construct nature.

Criticism on the Social Construction of Nature

The social construction of nature has room for improvement in four main areas:

- By giving more importance to how realities are culturally constructed through social interactions.
- By acknowledging that all science should be analyzed by the same standard.
- By gaining a better understanding of the role language plays in constructionism.
- By giving more importance to how truths exist and how they are justified, using Actor-Network Theory.

Language Geography

Language geography is the branch of human geography that studies the geographic distribution of languages or its constituent elements. Linguistic geography can also refer to studies of how people talk about the landscape. For example, toponymy is the study of place names. Landscape ethnoecology, also known as ethnophysiography, is the study of landscape ontologies and how they are expressed in language.

There are two principal fields of study within the geography of language:

- Geography of languages, which deals with the distribution through history and space of languages, and is concerned with 'the analysis of the distribution patterns and spatial structures of languages in contact'.
- Geolinguistics being, when used as a sub-discipline of geography, the study of the 'political, economic and cultural processes that affect the status and distribution of languages. When perceived as a sub-discipline of linguistics which incorporates contact linguistics, one definition appearing has been 'the study of languages and dialects in contact and in conflict with various societal, economic, ideological, political and other contemporary trends with regard to a particular geographic location and on a planetary scale'.

Various other terms and subdisciplines have been suggested, but none gained much currency, including:

- Linguistic geography, which deals with regional linguistic variations within languages, also called 'dialect geography' which some consider a subdivision of geolinguistics.
- A division within the examination of linguistic geography separating the studies of change over time and space.

Many studies in what is now called contact linguistics have researched the effect of language contact, as the languages or dialects (varieties) of peoples have interacted. This territorial expansion of language groups has usually resulted in the overlaying of languages upon existing speech areas, rather than the replacement of one language by another. An example could be sought in the Norman Conquest of England: Old French became the language of the aristocracy but Middle English remained the language of the majority of the population.

Linguistic Geography

Linguistic geography, as a field, is dominated by linguists rather than geographers. Charles W. J. Withers describes the difference as resulting from a focus on "elements of language, and only then with their geographical or social variation, as opposed to investigation of the processes making for change in the extent of language areas". Peter Trudgill says, "linguistic geography has been geographical only in the sense that it has been concerned with the spatial distribution of linguistic phenomena". Greater emphasis has been laid upon explanation rather than mere description of the patterns of linguistic change. That move has paralleled similar concerns in geography and language studies. Some studies have paid attention to the social use of language and to variations in dialect within languages in regard to social class or occupation. Regarding such variations, lexicographer Robert Burchfield notes that their nature "is a matter of perpetual discussion and disagreement" and notes that "most professional linguistic scholars regard it as axiomatic that all varieties of English have a sufficiently large vocabulary for the expression of all the distinctions that are important in the society using it". He contrasts this with the view of the historian John Vincent, who regards such a view as:

> A nasty little orthodoxy among the educational and linguistic establishment. However badly you need standard English, you will have the merits of non-standard English waved at you. The more extravagantly your disadvantages will be lauded as 'entirely adequate for the needs of their speakers', to cite the author of Sociolinguistics. It may sound like a radical cry to support pidgin, patois, or dialect, but translated into social terms, it looks more like a ploy to keep Them (whoever Them may be) out of the middle-class suburbs.
>
> — John Vincent, The Times

Burchfield concludes, "Resolution of such opposite views is not possible future of dialect studies and the study of class-marked distinctions are likely to be of considerable interest to everyone".

In England, linguistic geography has traditionally focused upon rural English, rather than urban English. A common production of linguistic investigators of dialects is the shaded and dotted map showing to show where one linguistic feature ends and another begins or overlaps.

Geolinguistic Organizations

Most geolinguistic organizations identify themselves as associations of linguists rather than of geographers. This includes the two oldest which both date to 1965 with "Amici Linguarum" (language friends) being founded by Erik V. Gunnemark and The American Society of Geolinguistics by Prof. Mario A. Pei. The research in geolinguistics which these organizations and others, which are more geographically oriented, promote is often interdisciplinary, being at times simultaneously both linguistic and geographic, and also being at times linked to other sub-disciplines of linguistics as well as going beyond linguistics to connect to sociology, anthropology, ethnology, history, demographics, political science, studies of cognition and communication, etc.

CULTURAL GEOGRAPHY

This map shows average hair colour by location in and around Europe.

Cultural geography is a subfield within human geography. Though the first traces of the study of different nations and cultures on Earth can be dated back to ancient geographers such as Ptolemy or Strabo, cultural geography as academic study firstly emerged as an alternative to the environmental determinist theories of the early 20th century, which had believed that people and societies are controlled by the environment in which they develop. Rather than studying pre-determined regions based upon environmental classifications, cultural geography became interested in cultural landscapes. This was led by the "father of cultural geography" Carl O. Sauer of the University of California, Berkeley. As a result, cultural geography was long dominated by American writers.

Geographers drawing on this tradition see cultures and societies as developing out of their local landscapes but also shaping those landscapes. This interaction between the natural landscape and humans creates the cultural landscape. This understanding is a foundation of cultural geography but has been augmented over the past forty years with more nuanced and complex concepts of culture, drawn from a wide range of disciplines including anthropology, sociology, literary theory, and feminism. No single definition of culture dominates within cultural geography. Regardless of their particular interpretation of culture, however, geographers wholeheartedly reject theories that treat culture as if it took place "on the head of a pin".

Some of the topics within the field of study are globalization has been theorised as an explanation for cultural convergence.

This geography studies the geography of culture:

- Theories of cultural hegemony or cultural assimilation via cultural imperialism.

- Cultural areal differentiation, as a study of differences in way of life encompassing ideas, attitudes, languages, practices, institutions and structures of power and whole range of cultural practices in geographical areas.

- Study of cultural landscapes and cultural ecology.

- Other topics include sense of place, colonialism, post-colonialism, internationalism, immigration, emigration and ecotourism.

Though the first traces of the study of different nations and cultures on Earth can be dated back to ancient geographers such as Ptolemy or Strabo, cultural geography as academic study firstly emerged as an alternative to the environmental determinist theories of the early Twentieth century, which had believed that people and societies are controlled by the environment in which they develop. Rather than studying pre-determined regions based upon environmental classifications, cultural geography became interested in cultural landscapes. This was led by Carl O. Sauer (called the father of cultural geography), at the University of California, Berkeley. As a result, cultural geography was long dominated by American writers.

Charles Booth in the 19th century produced a series of books with various maps highlighting poverty in the city.

Sauer defined the landscape as the defining unit of geographic study. He saw that cultures and societies both developed out of their landscape, but also shaped them too. This interaction between the natural landscape and humans creates the cultural landscape. Sauer's work was highly qualitative and descriptive and was challenged in the 1930s by the regional geography of Richard Hartshorne. Hartshorne called for systematic analysis of the elements that varied from place to place, a project taken up by the quantitative revolution. Cultural geography was sidelined by the positivist tendencies of this effort to make geography into a hard science although writers such as David Lowenthal continued to write about the more subjective, qualitative aspects of landscape.

In the 1970s, new kind of critique of positivism in geography directly challenged the deterministic and abstract ideas of quantitative geography. A revitalized cultural geography manifested itself in the engagement of geographers such as Yi-Fu Tuan and Edward Relph and Anne Buttimer with humanism, phenomenology, and hermeneutics. This break initiated a strong trend in human geography toward Post-positivism that developed under the label "new cultural geography" while deriving methods of systematic social and cultural critique from critical geography.

Ongoing Evolution of Cultural Geography

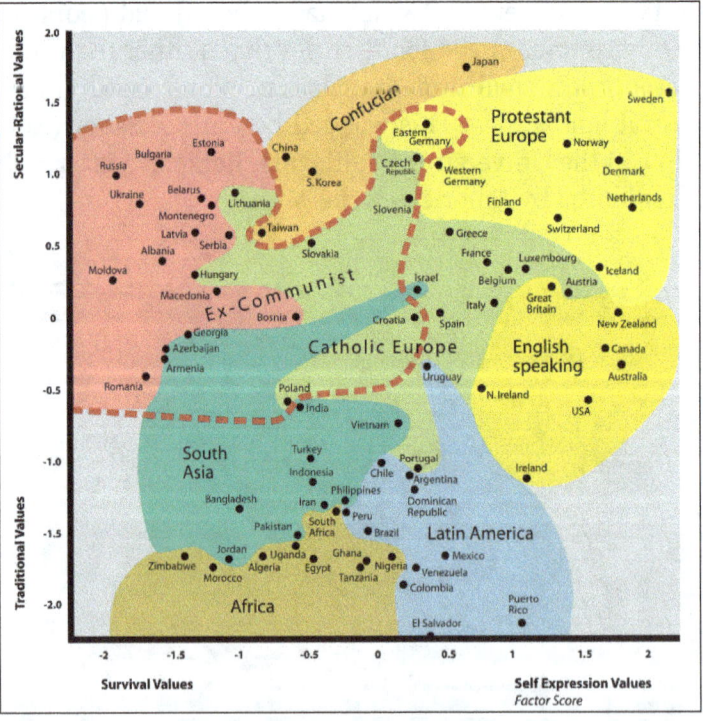

Cultural map of the world.

Since the 1980s, a "new cultural geography" has emerged, drawing on a diverse set of theoretical traditions, including Marxist political-economic models, feminist theory, post-colonial theory, post-structuralism and psychoanalysis.

Drawing particularly from the theories of Michel Foucault and performativity in western academia, and the more diverse influences of postcolonial theory, there has been a concerted effort to deconstruct the cultural in order to reveal that power relations are fundamental to spatial processes and sense of place. Particular areas of interest are how identity politics are organized in space and the construction of subjectivity in particular places.

Examples of areas of study include:

- Feminist geography,

- Children's geographies,

- Some parts of tourism geography,

- Behavioral geography,

- Sexuality and space,

- Some more recent developments in political geography,

- Music geography.

Some within the new cultural geography have turned their attention to critiquing some of its ideas, seeing its views on identity and space as static. It has followed the critiques of Foucault made by other 'poststructuralist' theorists such as Michel de Certeau and Gilles Deleuze. In this area, non-representational geography and population mobility research have dominated. Others have attempted to incorporate these and other critiques back into the new cultural geography.

Regional map of Gamelan, Kulintang, and Piphat music culture in Southeast Asia.

Groups within the geography community have differing views on the role of culture and how to analyze it in the context of geography. It is commonly thought that physical geography simply dictates aspects of culture such as shelter, clothing and cuisine. However, systematic development of this idea is generally discredited as environmental determinism. Geographers are now more likely to understand culture as a set of symbolic resources that help people make sense of the world around them, as well as a manifestation of the power relations between various groups and the structure through which social change is constrained and enabled. There are many ways to look at what culture means in light of various geographical insights, but in general geographers study how cultural processes involve spatial patterns and processes while requiring the existence and maintenance of particular kinds of places.

RELIGION AND GEOGRAPHY

Religion and geography is the study of the impact of geography, i.e. place and space, on religious belief.

Another aspect of the relationship between religion and geography is religious geography, in which geographical ideas are influenced by religion, such as early map-making, and the biblical geography that developed in the 16th century to identify places from the Bible.

Traditionally, the relationship between geography and religion can clearly be seen by the influences of religion in shaping cosmological understandings of the world. From the sixteenth and seventeenth century, the study of geography and religion mainly focused on mapping the spread of Christianity (termed ecclesiastical geography by Issac), though in the later half of the seventeenth century, the influences and spread of other religions were also taken into account.

Other traditional approaches to the study of the relationship between geography and religion involved the theological explorations of the workings of Nature – a highly environmentally deterministic approach which identified the role of geographical environments in determining the nature and evolution of different religious traditions.

Thus, geographers are less concerned about religion per se, but are more sensitive to how religion as a cultural feature affects social, cultural, political and environmental systems. The point of focus is not the specifics of religious beliefs and practices, but how these religious beliefs and practices are internalised by adherents, and how these processes of internalization influence, and is influenced by, social systems.

Sacred Places

Traditional cultural geographical approaches to the study of religion mainly seek to determine religion's impact on the landscape. A more contemporary approach to the study of the intersections of geography and religion not only highlights the role of religion in affecting landscape changes and in assigning sacred meanings to specific places, but also acknowledges how in turn, religious ideology and practice at specific spaces are guided and transformed by their location.

Religious experiences and the belief in religious meanings transforms physical spaces into sacred spaces. These perceptions and imaginings influence the way such spaces are used, and the personal, spiritual meanings developed in using such sacred spaces. These religiously significant spaces go beyond officially religious/spiritual spaces (such as places of worship) to include non-official religious spaces such as homes, schools and even bodies. These works have focused on both material aspects of spaces (such as architectural distinctiveness) and socially constructed spaces (such as rituals and demarcation of sacred spaces) to present religious meaning and significance.

A key focus in the study of sacred places is the politics of identity, belonging and meaning that are ascribed to sacred sites, and the constant negotiations for power and legitimacy. Particularly in multicultural settings, the contestation for legitimacy, public approval, and negotiations for use of particular spaces are at the heart of determining how communities understand, internalise and struggle to compete for the right to practice their religious traditions in public spaces.

Community and Identity

Religion may be a starting point to examine issues of ethnic identity formation and the construction of ethnic identity Geographers studying the negotiations of religious identity within various communities are often concerned with the overt articulation of religious identity, for example, how adherents in different locations establish their distinctive (religious and cultural) identities through their own understandings of the religion, and how they externally present their religious adherence (in terms of religious practice, ritual and behaviour). As an overarching theme, the articulation of religious identity is concerned with material aspects of symbolizing religious identity (such as architecture and the establishment of a physical presence), with negotiations and struggles in asserting religious identity in the face of persecution and exclusion and with personal practices of religious ritual and behaviour that re-establishes one's religious identity.

New Geographies of Religion

As research on geography and religion has grown, one of the new focuses of geographical research examines the rise of religious fundamentalism, and the resulting impact this has on the geographical contexts in which it develops.

In addition, migration processes have resulted in the development of religious pluralism in numerous countries, and the landscape changes that accompany the movement and settlement of communities defined by religion is a key focus in the study of geography and religion. More work needs to be done to examine the intersections and collisions that occur due to the movement of communities (for example, the migration of Muslim communities to western countries) and highlight how these communities negotiate their religious experiences in new spaces. Recent research in this area has been published by Barry A. Vann who analyzes Muslim population shifts in the Western world and the theological factors that play into these demographic trends.

Another new area of interest in the study of geography and religion explores different sites of religious practice beyond the 'officially sacred' – sites such as religious schools, media spaces, banking and financial practices (for example, Islamic banking) and home spaces are just some of the different avenues that take into account informal, everyday spaces that intersect with religious practice and meaning.

POLITICAL GEOGRAPHY

Political geography is concerned with the study of both the spatially uneven outcomes of political processes and the ways in which political processes are themselves affected by spatial structures. Conventionally, for the purposes of analysis, political geography adopts a three-scale structure with the study of the state at the centre, the study of international relations (or geopolitics) above it, and the study of localities below it. The primary concerns of the subdiscipline can be summarized as the inter-relationships between people, state, and territory.

The British geographer Halford Mackinder was also heavily influenced by environmental determinism and in developing his concept of the 'geographical pivot of history' or the Heartland Theory he argued that the era of sea power was coming to an end and that land based powers were in the ascendant, and, in particular, that whoever controlled the heartland of 'Euro-Asia' would control the world. This theory involved concepts diametrically opposed to the ideas of Alfred Thayer Mahan about the significance of sea power in world conflict. The heartland theory hypothesized the possibility of a huge empire being created which didn't need to use coastal or transoceanic transport to supply its military–industrial complex, and that this empire could not be defeated by the rest of the world allied against it. This perspective proved influential throughout the period of the Cold War, underpinning military thinking about the creation of buffer states between East and West in central Europe.

The heartland theory depicted a world divided into a Heartland (Eastern Europe/ Western Russia); World Island (Eurasia and Africa); Peripheral Islands (British Isles, Japan, Indonesia and Australia) and New World (The Americas). Mackinder argued that whoever controlled the Heartland would have control of the world. He used these ideas to politically influence events such as the Treaty of Versailles, where buffer states were created between the USSR and Germany, to prevent either of them controlling the Heartland. At the same time, Ratzel was creating a theory of states based around the concepts of Lebensraum and Social Darwinism. He argued that states were analogous to 'organisms' that needed sufficient room in which to live. Both of these writers created the idea of a political and geographical science, with an objective view of the world. Prior to World War II political geography was concerned largely with these issues of global power struggles and influencing state policy, and the above theories were taken on board by German geopoliticians such as Karl Haushofer who - perhaps inadvertently - greatly influenced Nazi political theory, which was a form of politics seen to be legitimated by such 'scientific' theories.

The close association with environmental determinism and the freezing of political boundaries during the Cold War led to a significant decline in the perceived importance of political geography, which was described by Brian Berry in 1968 as a 'moribund backwater'. Although at this time in most other areas of human geography new approaches, including quantitative spatial science, behavioural studies, and structural Marxism, were invigorating academic research these were largely ignored by political geographers whose main point of reference remained the regional approach. As a result, most of the political geography texts produced during this period were descriptive, and it was not until 1976 that Richard Muir could argue that political geography was no longer a dead duck, but could in fact be a phoenix.

The Brandenburg Gate of the Berlin Wall.

From the late-1970s onwards, political geography has undergone a renaissance, and could fairly be described as one of the most dynamic of the sub-disciplines today. The revival was underpinned by the launch of the journal Political Geography Quarterly (and its expansion to bi-monthly production as Political Geography). In part this growth has been associated with the adoption by political geographers of the approaches taken up earlier in other areas of human geography, for example, Ron J. Johnston's work on electoral geography relied heavily on the adoption of quantitative spatial science, Robert Sack's work on territoriality was based on the behavioural approach, Henry Bakis showed the impact of information and telecommunications networks on political geography, and Peter Taylor's work on World Systems Theory owed much to developments within structural Marxism. However, the recent growth in vitality and importance of this sub-discipline is also related to the changes in the world as a result of the end of the Cold War. With the emergence of a new world order (which as yet, is only poorly defined) and the development of new research agendas, such as the more recent focus on social movements and political struggles, going beyond the study of nationalism with its explicit territorial basis. There has also been increasing interest in the geography of green politics, including the geopolitics of environmental protest, and in the capacity of our existing state apparatus and wider political institutions, to address any contemporary and future environmental problems competently.

Political geography has extended the scope of traditional political science approaches by acknowledging that the exercise of power is not restricted to states and bureaucracies, but is part of everyday life. This has resulted in the concerns of political geography increasingly overlapping with those of other human geography sub-disciplines such as economic geography, and, particularly, with those of social and cultural geography in relation to the study of the politics of place. Although contemporary political geography maintains many of its traditional concerns the multi-disciplinary expansion into related areas is part of a general process within human geography which involves the blurring of boundaries between formerly discrete areas of study, and through which the discipline as a whole is enriched.

In particular, contemporary political geography often considers:

- How and why states are organized into regional groupings, both formally (e.g. the European Union) and informally (e.g. the Third World).

- The relationship between states and former colonies, and how these are propagated over time, for example through neo-colonialism.

- The relationship between a government and its people.

- The relationships between states including international trades and treaties.

- The functions, demarcations and policing of boundaries.

- How imagined geographies have political implications.

- The influence of political power on geographical space.

- The political implications of modern media (e.g. radio, TV, ICT, Internet, social networks).

- The study of election results (electoral geography).

Critical Political Geography

Critical political geography is mainly concerned with the criticism of traditional political geographies vis-a-vis modern trends. As with much of the move towards 'Critical geographies', the arguments have drawn largely from postmodern, post structural and postcolonial theories. Examples include:

- Feminist geography, which argues for recognition of the power relations as patriarchal and attempts to theorise alternative conceptions of identity and identity politics. Alongside related concerns such as Queer theory and Youth studies.

- Postcolonial theories which recognise the Imperialistic, universalising nature of much political geography, especially in Development geography.

- Environmental justice which addresses the fair treatment and meaningful involvement of all people regardless of race, color, or income with respect to the development, implementation, and enforcement of environmental laws, regulations, and policies. In other words, it is a human right for all people to share equally in the benefits bestowed by a healthy environment.

Geopolitics

Geopolitics is the study of the effects of Earth's geography (human and physical) on politics and international relations. While geopolitics usually refers to countries and relations between them, it may also focus on two other kinds of states: de facto independent states with limited international recognition and relations between sub-national geopolitical entities, such as the federated states that make up a federation, confederation or a quasi-federal system.

At the level of international relations, geopolitics is a method of studying foreign policy to understand, explain and predict international political behavior through geographical variables. These include area studies, climate, topography, demography, natural resources, and applied science of the region being evaluated.

Geopolitics focuses on political power linked to geographic space. In particular, territorial waters and land territory in correlation with diplomatic history. Topics of geopolitics include relations between the interests of international political actors and interests focused within an area, a space, or a geographical element; relations which create a geopolitical system. "Critical geopolitics" deconstructs classical geopolitical theories, by showing their political/ideological functions for great powers. Recently, there are some works which discuss the geopolitics of renewable energy.

According to Christopher Gogwilt and other researchers, the term is currently being used to describe a broad spectrum of concepts, in a general sense used as "a synonym for international political relations", but more specifically "to imply the global structure of such relations", which builds on "early-twentieth-century term for a pseudoscience of political geography" and other pseudoscientific theories of historical and geographic determinism.

Alfred Thayer Mahan , a frequent commentator on world naval strategic and diplomatic affairs, believed that national greatness was inextricably associated with the sea— and particularly with its commercial use in peace and its control in war. Mahan's theoretical framework came from Antoine-Henri Jomini, and emphasized that strategic locations (such as chokepoints, canals, and coaling stations), as well as quantifiable levels of fighting power in a fleet, were conducive to control over the sea. He proposed six conditions required for a nation to have sea power:

- Advantageous geographical position;

- Serviceable coastlines, abundant natural resources, and favorable climate;

- Extent of territory;

- Population large enough to defend its territory;

- Society with an aptitude for the sea and commercial enterprise; and

- Government with the influence and inclination to dominate the sea.

Mahan distinguished a key region of the world in the Eurasian context, namely, the Central Zone of Asia lying between 30° and 40° north and stretching from Asia Minor to Japan. In this zone independent countries still survived – Turkey, Persia, Afghanistan, China, and Japan. Mahan regarded those countries, located between Britain and Russia, as if between "Scylla and Charybdis". Of the two monsters – Britain and Russia – it was the latter that Mahan considered more threatening to the fate of Central Asia. Mahan was impressed by Russia's transcontinental size and strategically favorable position for southward expansion. Therefore, he found it necessary for the Anglo-Saxon "sea power" to resist Russia.

Homer Lea in The Day of the Saxon described that the entire Anglo-Saxon race faced a threat from German (Teuton), Russian (Slav), and Japanese expansionism: The "fatal" relationship of Russia, Japan, and Germany "has now assumed through the urgency of natural forces a coalition directed against the survival of Saxon supremacy". It is "a dreadful Dreibund". Lea believed that while Japan moved against Far East and Russia against India, the Germans would strike at England, the center of the British Empire. He thought the Anglo-Saxons faced certain disaster from their militant opponents.

Two famous Security Advisors from the cold war period, Henry Kissinger and Zbigniew Brzezinski, argued to continue the United States geopolitical focus on Eurasia and, particularly on Russia, despite the dissolution of the USSR and the end of the Cold War. Both continued their influence on geopolitics after the end of the Cold War, writing books on the subject in the 1990s—Diplomacy and The Grand Chessboard: American

Primacy and Its Geostrategic Imperatives. The Anglo-American classical geopolitical theories were revived.

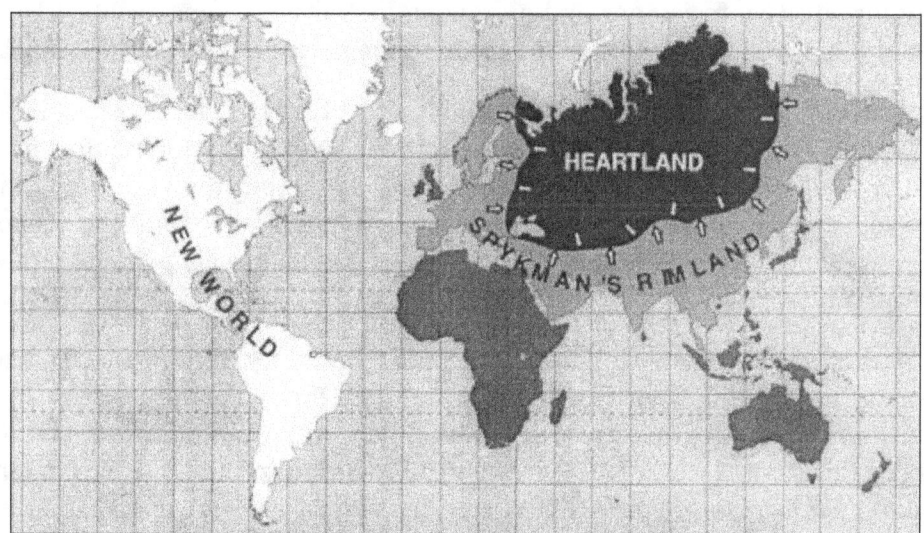

World map with the concepts of Heartland and Rimland applied.

Kissinger argued against the belief that with the dissolution of the USSR, hostile intentions had come to an end and traditional foreign policy considerations no longer applied. "They would argue that Russia, regardless of who govern it, sits astride the territory Halford Mackinder called the geopolitical heartland, and is the heir to one of the most potent imperial traditions". Therefore the United States must "maintain the global balance of power vis-à-vis the country with a long history of expansionism".

After Russia, the second geopolitical threat remained was Germany and, as Mackinder had feared ninety years ago, its partnership with Russia. During the Cold War, Kissinger argues, both sides of the Atlantic recognized that, "unless America is organically involved in Europe, it would be obliged to involve itself later under circumstances far less favorable to both sides of the Atlantic. That is even more true today. Germany has become so strong that existing European institutions cannot by themselves strike a balance between Germany and its European partners. Nor can Europe, even with Germany, manage by itself Russia". Thus Kissinger belied it is in no country's interest that Germany and Russia should fixate on each other as a principal partner. They would raise fears of condominium. Without America, Britain and France cannot cope with Germany and Russia; and "without Europe, America could turn into an island off the shores of Eurasia".

Spykman's vision of Eurasia was strongly confirmed: "Geopolitically, America is an island off the shores of the large landmass of Eurasia, whose resources and population far exceed those of the United States. The domination by a single power of either of Eurasia's two principal spheres—Europe and Asia—remains a good definition of strategic danger for America. Cold War or no Cold War. For such a grouping would have the capacity to outstrip America economically and, in the end, militarily. That danger would have to be resisted even were the dominant power apparently benevolent, for if the intentions ever changed, America would find itself with a grossly diminished capacity for effective resistance and a growing inability to shape events". The main interest of the American leaders is maintaining the balance of power in Eurasia.

Having converted from ideologist into geopolitician, Kissinger in retrospect interpreted the Cold War in geopolitical terms—an approach not characteristic for his works during the Cold War. Now, however, he stressed on the beginning of the Cold War: "The objective of moral opposition to Communism had merged with the geopolitical task of containing the Soviet expansion". Nixon, he added, was geopolitical rather than ideological cold warrior.

Three years after Kissinger's Diplomacy, Zbigniew Brzezinski followed suit, launching The Grand Chessboard: American Primacy and Its Geostrategic Imperatives and, after three more years, The Geostrategic Triad: Living with China, Europe, and Russia. The Grand Chessboard described the American triumph in the Cold War in terms of control over Eurasia: for the first time ever, a "non-Eurasian" power had emerged as a key arbiter of "Eurasian" power relations. The book states its purpose: "The formulation of a comprehensive and integrated Eurasian geostrategy is therefore the purpose of this book". Although the power configuration underwent a revolutionary change, Brzezinski confirmed three years later, Eurasia was still a megacontinent. Like Spykman, Brzezinski acknowledges that: "Cumulatively, Eurasia's power vastly overshadows America's".

In classical Spykman terms, Brzezinski formulized his geostrategic "chessboard" doctrine of Eurasia, which aims to prevent the unification of this megacontinent.

> "Europe and Asia are politically and economically powerful. It follows that American foreign policy must employ its influence in Eurasia in a manner that creates a stable continental equilibrium, with the United States as the political arbiter. Eurasia is thus the chessboard on which the struggle for global primacy continues to be played, and that struggle involves geo- strategy – the strategic management of geopolitical interests. But in the meantime it is imperative that no Eurasian challenger emerges, capable of dominating Eurasia and thus also of challenging America. For America the chief geopolitical prize is Eurasia and America's global primacy is directly dependent on how long and how effectively its preponderance on the Eurasian continent is sustained".

The Austro-Hungarian historian Emil Reich is considered to be the first having coined the acceptance in English as early as 1902 and later published in England in 1904.

Sir Halford Mackinder's Heartland Theory initially received little attention outside geography, but some thinkers would claim that it subsequently influenced the foreign policies of world powers. Those scholars who look to MacKinder through critical lenses accept him as an organic strategist who tried to build a foreign policy vision for Britain with his Eurocentric analysis of historical geography. His formulation of the Heartland Theory was set out in his article entitled "The Geographical Pivot of History", published in England in 1904. Mackinder's doctrine of geopolitics involved concepts diametrically opposed to the notion of Alfred Thayer Mahan about the significance of navies (he coined the term sea power) in world conflict. He saw navy as a basis of Colombian era empire, and predicted the 20th century to be domain of land power. The Heartland theory hypothesized a huge empire being brought into existence in the Heartland—which wouldn't need to use coastal or transoceanic transport to remain coherent. The basic notions of Mackinder's doctrine involve considering the geography of the Earth as being divided into two sections: the World Island or Core, comprising Eurasia and Africa; and the Peripheral "islands", including the Americas, Australia, Japan, the British Isles, and Oceania. Not only was the Periphery noticeably smaller than the World Island, it necessarily required much sea transport to function at the technological level of the World Island—which contained sufficient natural resources for a developed economy.

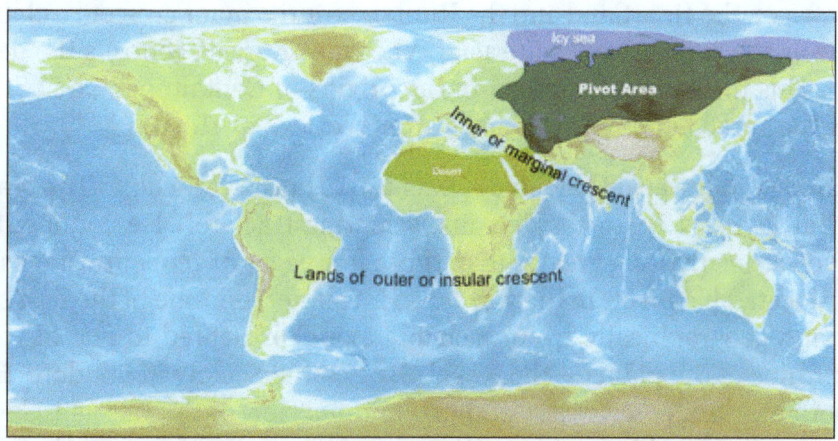

Sir Halford Mackinder's Heartland concept showing the situation of the
"pivot area" established in the Theory of the Heartland. He later revised it to mark
Northern Eurasia as a pivot while keeping area marked above as Heartland.

Mackinder posited that the industrial centers of the Periphery were necessarily located in widely separated locations. The World Island could send its navy to destroy each one of them in turn, and could locate its own industries in a region further inland than the Periphery (so they would have a longer struggle reaching them, and would face a well-stocked industrial bastion). Mackinder called this region the Heartland. It essentially comprised Central and Eastern Europe: Ukraine, Western Russia, and Mitteleuropa. The Heartland contained the grain reserves of Ukraine, and many other natural resources. Mackinder's notion of geopolitics was summed up when he said:

> "Who rules Central and Eastern Europe commands the Heartland. Who rules the Heartland commands the World-Island. Who rules the World-Island commands the World."

Nicholas J. Spykman was both a follower and critic of geostrategists Alfred Mahan, and Halford Mackinder. His work was based on assumptions similar to Mackinder's, including the unity of world politics and the world sea. He extends this to include the unity of the air. Spykman adopts Mackinder's divisions of the world, renaming some:

- The Heartland;

- The Rimland (analogous to Mackinder's "inner or marginal crescent" also an intermediate region, lying between the Heartland and the marginal sea powers);

- The Offshore Islands & Continents (Mackinder's "outer or insular crescent").

Under Spykman's theory, a Rimland separates the Heartland from ports that are usable throughout the year (that is, not frozen up during winter). Spykman suggested this required that attempts by Heartland nations (particularly Russia) to conquer ports in the Rimland must be prevented. Spykman modified Mackinder's formula on the relationship between the Heartland and the Rimland (or the inner crescent), claiming that "Who controls the rimland rules Eurasia. Who rules Eurasia controls the destinies of the world". This theory can be traced in the origins of Containment, a U.S. policy on preventing the spread of Soviet influence after World War II.

Another famous follower of Mackinder was Karl Haushofer who called Mackinder's Geographical Pivot of History a "genius' scientific tractate". He commented on it: "Never have I seen anything greater than those few pages of geopolitical masterwork". Mackinder located his Pivot, in the words of Haushofer, on "one of the first solid, geopolitically and geographically irreproachable maps, presented to one of the earliest scientific forums of the planet – the Royal Geographic Society in London" Haushofer adopted both Mackinder's Heartland thesis and his view of the Russian-German alliance – powers that Mackinder saw as the major contenders for control of Eurasia in the twentieth century. Following Mackinder he suggested an alliance with the Soviet Union and, advancing a step beyond Mackinder, added Japan to his design of the Eurasian Bloc.

In 2004, at the centenary of The Geographical Pivot of History, famous Historian Paul Kennedy wrote: "Right now with hundreds of thousands of US troops in the Eurasian rimlands and with administration constantly explaining why it has to stay the course, it looks as if Washington is taking seriously Mackinder's injunction to ensure control of the geographical pivot of history".

Friedrich Ratzel influenced by thinkers such as Darwin and zoologist Ernst Heinrich Haeckel, contributed to 'Geopolitik' by the expansion on the biological conception of geography, without a static conception of borders. Positing that states are organic and growing, with borders representing only a temporary stop in their movement, he held that the expanse of a state's borders is a reflection of the health of the nation—meaning that static countries are in decline. Ratzel published several papers, among which was the essay "Lebensraum" concerning biogeography. Ratzel created a foundation for the German variant of geopolitics, geopolitik. Influenced by the American geostrategist Alfred Thayer Mahan, Ratzel wrote of aspirations for German naval reach, agreeing that sea power was self-sustaining, as the profit from trade would pay for the merchant marine, unlike land power.

The geopolitical theory of Ratzel has been criticized as being too sweeping, and his interpretation of human history and geography being too simple and mechanistic. Critically, he also underestimated the importance of social organization in the development of power.

After World War I, the thoughts of Rudolf Kjellén and Ratzel were picked up and extended by a number of German authors such as Karl Haushofer , Erich Obst, Hermann Lautensach and Otto Maull. In 1923, Karl Haushofer founded the Zeitschrift für Geopolitik, which was later used in the propaganda of Nazi Germany. The key concepts of Haushofer's Geopolitik were Lebensraum, autarky, pan-regions, and organic borders. States have, Haushofer argued, an undeniable right to seek natural borders which would guarantee autarky.

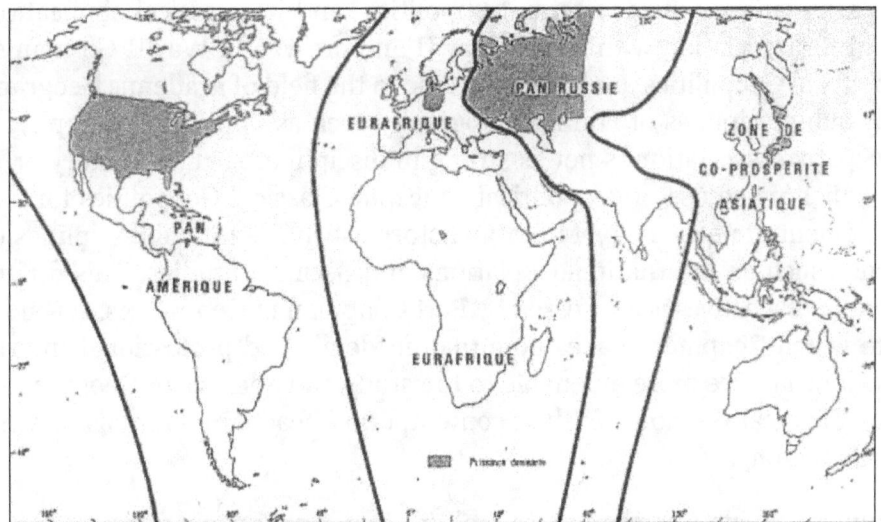

Division of the world according to Haushofer's Pan-Regions Doctrine.

Haushofer's influence within the Nazi Party has recently been challenged, given that Haushofer failed to incorporate the Nazis' racial ideology into his work. Popular views of the role of geopolitics in the Nazi Third Reich suggest a fundamental significance on the part of the geo-politicians in the ideological orientation of the Nazi state. Bassin reveals that these popular views are in important ways misleading and incorrect.

Despite the numerous similarities and affinities between the two doctrines, geopolitics was always held suspect by the National Socialist ideologists. This was understandable, for the underlying philosophical orientation of geopolitics did not comply with that of National Socialism. Geopolitics shared Ratzel's scientific materialism and geographic determinism, and held that human society was determined by external influences—in the face of which qualities held innately by individuals or groups were of reduced or no significance. National Socialism rejected in principle both materialism and determinism and also elevated innate human qualities, in the form of a hypothesized 'racial character,' to the factor of greatest significance in the constitution of human society. These differences led after 1933 to friction and ultimately to open denunciation of geopolitics by Nazi ideologues. Nevertheless, German Geopolitik was discredited by its (mis)use in Nazi expansionist policy of World War II and has never achieved standing comparable to the pre-war period.

The resultant negative association, particularly in U.S. academic circles, between classical geopolitics and Nazi or imperialist ideology, is based on loose justifications. This has been observed in particular by critics of contemporary academic geography, and proponents of a "neo"-classical geopolitics in particular. These include Haverluk et al., who argue that the stigmatization of geopolitics in academia is unhelpful as geopolitics as a field of positivist inquiry maintains potential in researching and resolving topical, often politicized issues such as conflict resolution and prevention, and mitigating climate change.

Negative associations with the term "geopolitics" and its practical application stemming from its association with World War II and pre-World War II German scholars and students of Geopolitics are largely specific to the field of academic Geography, and especially sub-disciplines of Human Geography such as Political Geography. However, this negative association is not as strong in disciplines such as History or Political Science, which make use of geopolitical concepts. Classical Geopolitics forms an important element of analysis for Military History as well as for subdisciplines of Political Science such as International Relations and Security Studies. This difference in disciplinary perspectives is addressed by Bert Chapman in Geopolitics: A Guide To the Issues, in which Chapman makes note that academic and professional International Relations journals are more amenable to the study and analysis of Geopolitics, and in particular Classical Geopolitics, than contemporary academic journals in the field of Political Geography.

In disciplines outside Geography, Geopolitics is not negatively viewed (as it often is among academic geographers such as Carolyn Gallaher or Klaus Dodds) as a tool of Imperialism or associated with Nazism, but rather viewed as a valid and consistent manner of assessing major international geopolitical circumstances and events, not necessarily related to armed conflict or military operations.

French geopolitical doctrines broadly opposed to German Geopolitik and reject the idea of a fixed geography. French geography is focused on the evolution of polymorphic territories being the result of mankind's actions. It also relies on the consideration of long time periods through a refusal to take specific events into account. This method has been theorized by Professor Lacoste according to three principles: Representation; Diachronie; and Diatopie.

In The Spirit of the Laws, Montesquieu outlined the view that man and societies are influenced by climate. He believed that hotter climates create hot-tempered people and colder climates aloof people, whereas the mild climate of France is ideal for political systems. Considered as one of the founders of French geopolitics, Élisée Reclus, is the author of a book considered as a reference in modern geography (Nouvelle Géographie universelle). Alike Ratzel, he considers geography through a global vision. However, in complete opposition to Ratzel's vision, Reclus considers geography not to be unchanging; it is supposed to evolve commensurately to the development of human society. His marginal political views resulted in his rejection by academia.

French geographer and geopolitician Jacques Ancel is considered to be the first theoretician of geopolitics in France, and gave a notable series of lectures at the European Center of the Carnegie Endowment for International Peace in Paris and published Géopolitique in 1936. Like Reclus, Ancel rejects German determinist views on geopolitics (including Haushofer's doctrines).

Braudel's broad view used insights from other social sciences, employed the concept of the longue durée, and downplayed the importance of specific events. This method was inspired by the French geographer Paul Vidal de la Blache (who in turn was influenced by German thought, particularly that of Friedrich Ratzel whom he had met in Germany). Braudel's method was to analyse the interdependence between individuals and their environment. Vidalian geopolitics is based on varied forms of cartography and on possibilism (founded on a societal approach of geography—i.e. on the principle of spaces polymorphic faces depending from many factors among them mankind, culture, and ideas) as opposed to determinism.

Due to the influence of German Geopolitik on French geopolitics, the latter were for a long time banished from academic works. In the mid-1970s, Yves Lacoste—a French geographer who was directly inspired by Ancel, Braudel and Vidal de la Blache—wrote La géographie, ça sert d'abord à faire la guerre (Geography first use is war) in 1976. This book—symbolizes the birth of this new school of geopolitics. Initially linked with communist party evolved to a less liberal approach. At the end of the 1980s he founded the Institut Français de Géopolitique (French Institute for Geopolitics) that publishes the Hérodote revue. While rejecting the generalizations and broad abstractions employed by the German and Anglo-American traditions (and the new geographers), this school does focus on spatial dimension of geopolitics affairs on different levels of analysis. This approach emphazises the importance of multi-level (or multi-scales) analysis and maps at the opposite of critical geopolitics which avoid such tools. Lacoste proposed that every conflict (both local or global) can be considered from a perspective grounded in three assumptions:

- Representation: Each group or individuals is the product of an education and is characterized by specific representations of the world or others groups or individuals. Thus, basic societal beliefs are grounded in their ethnicity or specific location. The study of representation is a common point with the more contemporary critical geopolitics. Both are connected with the work of Henri Lefebvre.

- Diachronie: Conducting an historical analysis confronting "long periods" and short periods as the prominent French historian Fernand Braudel suggested.

- Diatopie: Conducting a cartographic survey through a multiscale mapping.

Connected with this stream, and former member of Hérodote editorial board, the French geographer Michel Foucher developed a long term analysis of international borders. He coined various neologism among them: Horogenesis: Neologism that describes the concept of studying the birth of borders, Dyade: border shared by two neighbouring states (for instance US territory has two terrestrial dyades: one with Canada and one with Mexico). The main book of this searcher "Fronts et frontières" (Fronts and borders) first published in 1991, without equivalent remains as of yet untranslated in English. Michel Foucher is an expert of the African Union for borders affairs.

More or less connected with this school, Stéphane Rosière can be quoted as the editor in Chief of the online journal L'Espace politique, this journal created in 2007 became the most prominent French journal of political geography and Geopolitics with Hérodote.

A much more conservative stream is personified by François Thual. Thual was a French expert in geopolitics, and a former official of the Ministry of Civil Defence. Thual taught geopolitics of the religions at the French War College, and has written thirty books devoted mainly to geopolitical method and its application to various parts of the world. He is particularly interested in the Orthodox, Shiite, and Buddhist religions, and in troubled regions like the Caucasus. Connected with F. Thual, Aymeric Chauprade, former professor of geopolitics at the French War College and now member of the extreme-right party "Front national", subscribes to a supposed "new" French school of geopolitics which advocates above all a return to realpolitik and "clash of civilization" (Huntington). The thought of this school is expressed through the French Review of Geopolitics (headed by Chauprade) and the International Academy of Geopolitics. Chauprade is a supporter of a Europe of nations, he advocates a European Union excluding Turkey, and a policy of compromise with Russia (in the frame of a Eurasian alliance which is en vogue among European extreme-right politists) and supports the idea of a multipolar world—including a balanced relationship between China and the U.S.

French philosopher Michel Foucault's dispositif introduced for the purpose of biopolitical research was also adopted in the field of geopolitical thought where it now plays a central role.

In the 1990s a senior researcher at the Institute of Philosophy, Russian Academy of Sciences of the Russian Academy of Sciences, Vadim Tsymbursky, coined the term "island-Russia" and developed the "Great Limitrophe" concept.

Colonel-General Leonid Ivashov (retired), a Russian geopolitics specialist of the early 21st century, headed the Academy of Geopolitical Problems, which analyzes the international and domestic situations and develops geopolitical doctrine. Earlier, Colonel-General Leonid Ivashov headed the Main Directorate for International Military Cooperation of the Ministry of Defence of the Russian Federation.

Vladimir Karyakin, leading researcher at the Russian Institute for Strategic Studies, has proposed the term "geopolitics of the third wave".

Aleksandr Dugin, a Russian fascist and nationalist who has developed a close relationship with Russia's Academy of the General Staff wrote "The Foundations of Geopolitics: The Geopolitical Future of Russia" in 1997, which has had a large influence within the Russian military, police, and foreign policy elites and it has been used as a textbook in the Academy of the General Staff of the Russian military. Its publication in 1997 was well-received in Russia and powerful Russian political figures subsequently took an interest in Dugin.

URBAN GEOGRAPHY

Urban geography is the subdiscipline of geography that derives from a study of cities and urban processes. Urban geographers and urbanists examine various aspects of urban life and the built environment. Scholars, activists, and the public have participated in, studied, and critiqued flows of economic and natural resources, human and non-human bodies, patterns of development and infrastructure, political and institutional activities, governance, decay and renewal, and notions of socio-spatial inclusions, exclusions, and everyday life.

Urban geographers are primarily concerned with the ways in which cities and towns are constructed, governed and experienced. Alongside neighboring disciplines such as urban anthropology, urban planning and urban sociology, urban geography mostly investigates the impact of urban processes on the earth's surface's social and physical structures. Urban geographical research can be part of both human geography and physical geography.

The two fundamental aspects of cities and towns, from the geographic perspective are:

- Location ("systems of cities"): Spatial distribution and the complex patterns of movement, flows and linkages that bind them in space.

- Urban structure ("cities as systems"): Study of patterns of distribution and interaction within cities, from quantitative, qualitative, structural, and behavioral perspectives.

Cities as Centers of Manufacturing and Services

Cities differ in their economic makeup, their social and demographic characteristics, and the roles they play within the city system. One can trace these differences back to regional variations in the local resources on which growth was based during the early development of the urban pattern and in part to the subsequent shifts in the competitive advantage of regions brought about by changing locational forces affecting regional specialization within the framework of a market economy. The recognition of different city types is critical for the classification of cities in urban geography. For such classification, emphasis given in particular to functional town classification and the basic underlying dimensions of the city system.

The purpose of classifying cities is twofold. On the one hand, it is undertaken to search reality for hypotheses. In this context, the recognition of different types of cities on the basis of, for example, their functional specialization may enable the identification of spatial regularities in the distribution and structure of urban functions and the formulation of hypotheses about the resulting patterns. On the other hand, classification is undertaken to structure reality in order to test specific hypotheses that have already been formulated. For example, to test the hypotheses that cities with a diversified economy grow at a faster rate then those with a more specialized economic base, cities must first be classified so that diversified and specialized cities can be differentiated.

The simplest way to classify cities is to identify the distinctive role they play in the city system. There are three distinct roles:

- Central places functioning primarily as service centers for local hinterlands.

- Transportation cities performing break-of-bulk and allied functions for larger regions.

- Specialized-function cities, dominated by one activity such as mining, manufacturing or recreation and serving national and international markets.

The composition of a city's labor force has traditionally been regarded as the best indicator of functional specialization, and different city types have been most frequently identified from the analysis of employment profiles. Specialization in a given activity is said to exist when employment in it exceeds some critical level.

The relationship between the city system and the development of manufacturing has become very apparent. The rapid growth and spread of cities within the heartland-hinterland framework after 1870 was conditioned to a large extent by industrial developments, and the decentralization of population within the urban system in recent years is related in large part to the movement of employment in manufacturing away from traditional industrial centers. Manufacturing is found in nearly all cities, but its importance is measured by the proportion of total earnings received by the inhabitants of an urban area. When 25 percent or more of the total earnings in an urban region derive from manufacturing, that urban area is arbitrarily designated as a manufacturing center.

The location of manufacturing is affected by myriad economic and non-economic factors, such as the nature of the material inputs, the factors of production, the market and transportation costs. Other important influences include agglomeration and external economies, public policy and personal preferences. Although it is difficult to evaluate precisely the effect of the market on the location of manufacturing activities, two considerations are involved:

- The nature of and demand for the product.
- Transportation costs.

Urbanization

Urbanization, the transformation of population from rural to urban, is a major phenomenon of the modern era and a central topic of study.

Urban geography arrived as a critical sub-discipline with the 1973 publication of David Harvey's Social Justice and the City, which was heavily influenced by previous work by Anne Buttimer. Prior to its emergence as its own discipline, urban geography served as the academic extension of what was otherwise a professional development and planning practice. At the turn of the 19th century, urban planning began as a profession charged with mitigating the negative consequences of industrialization as documented by Friedrich Engels in his geographic analysis of the condition of the working class in England, 1844.

In a 1924 study of urban geography, Marcel Aurousseau observed that urban geography cannot be considered a subdivision of geography because it plays such an important part. However, urban geography did emerge as a specialized discipline after World War II, amidst increasing urban planning and a shift away from the primacy of physical terrain in the study of geography. Chauncy Harris and Edward Ullman were among its earliest exponents.

Urban geography arose by the 1930s in the Soviet Union as an academic complement to active urbanization and communist urban planning, focusing on cities' economic roles and potential.

Spatial analysis, behavioral analysis, Marxism, humanism, social theory, feminism, and postmodernism have arisen (in approximately this order) as overlapping lenses used within the field of urban geography in the West.

Geographic information science, using digital processing of large data sets, has become widely used since the 1980s, with major applications for urban geography.

TRANSPORT GEOGRAPHY

Transport geography, also transportation geography, is a branch of geography that investigates the movement and connections between people, goods and information on the Earth's surface.

Spatial interaction.

Transportation geography detects, describes, and explains the Earth's surface's transportation spaces regarding location, substance, form, function, and genesis. It also investigates the effects of transportation on land use, on the physical material patterns at the surface of the earth known as 'cover patterns', and on other spatial processes such as environmental alterations. Moreover, it contributes to transport, urban, and regional planning.

Transportation is fundamental to the economic activity of exchange. Therefore, transport geography and economic geography are largely interrelated. At the most basic level, humans move and thus interact with each other by walking, but transportation geography typically studies more complex regional or global systems of transportation that include multiple interconnected modes like public transit, personal cars, bicycles, freight railroads, the Internet, airplanes and more. Such systems are increasingly urban in character. Thus, transport and urban geography are closely intertwined. Cities are very much shaped, indeed created, by the types of exchange and interaction facilitated by movement. Increasingly since the 19th century, transportation is seen as a way cities, countries or firms compete with each other in a variety of spaces and contexts.

Transportation Modes

In terms of transport modes, the primary forms are air, road, water, and rail. Each form has its own cost associated with 'speed of movement', which is affected by friction, place of origin, and destination. Ships are generally used for moving large amounts of goods. Maritime shipping is able to carry more around the world at a cheaper cost. For moving people who prefer to minimize travel time and maximize comfort and convenience, road and air are the most common modes in use. A railroad is often used to transport goods in areas away from water. Railroads may also be the source of transportation for people as well.

> " Transportation modes are an essential component of transport systems since they are the means by which mobility is supported. Geographers consider a wide range of modes that may be grouped into three broad categories based on the medium they exploit: land, water and air. Each mode has its own requirements and features, and is adapted to serve the specific demands of freight and passenger traffic. This gives rise to marked differences in the ways the modes are deployed and used in different parts of the world. Recently, there is a trend towards integrating the modes through intermodality and linking the modes ever more closely into production and distribution activities. At the same time; however, passenger and freight activity is becoming increasingly separated across most modes".

Road Transportation

Road transportation networks are connected with movements on constructed roads; carrying people and goods from one place to another by means of lorries, cars, etc. Transportation may be further categorized by the vehicle used or the purpose for transport itself.

Traffic jam.

Maritime Transportation

Water transportation is the slowest form of transportation in the movement of goods and people. Strategic chokepoints around the world have continued to play significant roles in maritime industry. Although the slowest form of transportation compared to road and rail transport, it is the most cost effective.

Rail Transportation

Rail transportation.

Rail transportation is the movement of cargo, goods, and passengers on trains as a form of transportation. Transportation by rails has been established as one of the safest modes of transportation over time.

Challenges for Transportation

Transportation availability on existing streets, highways, and rail facilities no longer match the transportation demands created by subsequent population growth and new location patterns of economic activity. Besides an increase in population, another problem is vehicles overloading the network of highways and arterial streets.

The well-being of poor people and people who live in developing areas can be threatened by systems of transportation that fail to connect them to jobs and medical assistance. For example, areas of Southern California have transportation systems that do not connect the homeless to these necessities.

TOURISM GEOGRAPHY

The content of tourism geography is complex, making a connection between the two concepts of geography and tourism, being rather new compared especially with the term of geography. The beginning of the science can be traced at the beginning of the 20th century, although tourism was being used inside the study of geography long before. By the 1950s, tourism geography began to be accepted as its own domain, especially in scientific works from USA and Germany. The first definitions were pretty vague and incomplete, G. Chabot stating that geography and tourism are two terms predestined to be joined because every geographer has to necessarily be doubled by the qualities of a tourist and also reciprocally, we can say that in every tourist there is a hidden geographer, because the intelligent tourist is actually a geographer that has not discovered himself. As more and more researchers began to study this new field, the accuracy and depth of the definitions began to improve.

The Role of Tourism Geography

As the importance and popularity of tourism increased, especially in the last two or three decades, becoming one of the biggest industries in the world, so did the role of tourism in geography and its study. While before there were few mentions of tourism related facts in any book or research of geography, today we cannot imagine any geographical descriptions without a separate chapter on tourism. Still rather raw and simple, L. Merlo considers this science as being a branch of geography that studies the position and appearance of tourist centers, their individual natural and cultural-historical characteristics, the attractions and traditions in the context of the area where they are found, the transportation network assuring the accessibility and the links with other tourist centers. Tourism is essentially a geographical phenomenon, regarding the transfer of people and services through space and time, so a special domain dedicated to the research of the interconnections between tourism and geography was inevitable. Although the scientific field is new, the connections of geography and travel can be traced to ancient times, when geographers had no other way of describing the world than traveling and seeing it for themselves.

The Connect between Tourism and Geography

The connections between tourism and geography are linked to specific terms such as place, location, space, accessibility, scale and others. This science also has an integrative character, containing key elements from all fields of geography, physical, human and economic. Besides this, tourism geography also has many common points with other sciences, including history, geology, biology, art, economy and so on. In more modern times, the tourism geography has become to achieve a broader definition, regarding the study of the spatial and temporal genesis, repartition and unfolding of the tourism phenomenon, being considered as a complex and specific interaction at the level of the geographic environment. As such, tourism geography studies things like the tourist resources (natural or man-made), the tourism infrastructure (transportation, accommodation, etc.), the types and forms of tourism, the tourist circulation (statistical research), tourist markets, as well as other domains. The areas of geographical interest in tourism are stated by S. Williams including the effect of scale, spatial distributions of tourist phenomena, tourism impacts, planning for tourism and spatial modeling of tourism development.

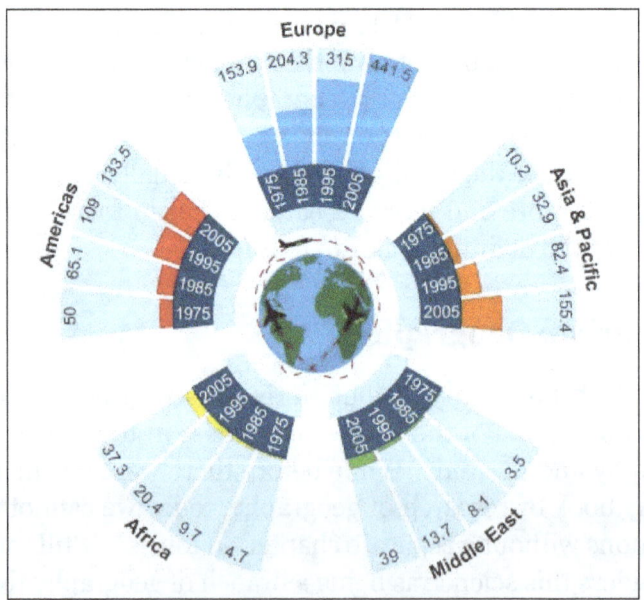

Increase in world tourism.

There is also another link between the two domains, as the primary factor which attracts tourists to a certain area is geography, with all its specific elements. The interconnections go a lot deeper, as tourists usually choose a certain destination primarily through the perceived experience of that place, as they envision its geographical characteristics, they use means of transportation to travel over the land or water surface, creating what we call tourism fluxes or the tourist circulation. While visiting a certain place, tourists actively discover and appreciate the geography of that place, from the landscapes with their typical forms, to the traditions of the local population, all while benefiting the local economy and using its resources. In conclusion, tourism geography studies the relations between places, landscapes and people, describing travel and tourism as an economic, social and cultural activity. More concisely, it is all about the spatial and temporal dynamics, as well as the interactions between the tourism resources.

ECONOMIC GEOGRAPHY

Economic geography is the subfield of human geography which studies economic activity. It can also be considered a subfield or method in economics.

Economic geography takes a variety of approaches to many different topics, including the location of industries, economies of agglomeration (also known as "linkages"), transportation, international trade, development, real estate, gentrification, ethnic economies, gendered economies, core-periphery theory, the economics of urban form, the relationship between the environment and the economy (tying into a long history of geographers studying culture-environment interaction), and globalization.

There are varied methodological approaches. Neoclassical location theorists, following in the tradition of Alfred Weber, tend to focus on industrial location and use quantitative methods. Since the 1970s, two broad reactions against neoclassical approaches have significantly changed the discipline: Marxist political economy, growing out of the work of David Harvey; and the new economic geography which takes into account social, cultural, and institutional factors in the spatial economy.

Economists such as Paul Krugman and Jeffrey Sachs have also analyzed many traits related to economic geography. Krugman called his application of spatial thinking to international trade theory the "new economic geography", which directly competes with an approach within the discipline of geography that is also called "new economic geography". The name geographical economics has been suggested as an alternative.

Early approaches to economic geography are found in the seven Chinese maps of the State of Qin, which date to the 4th century BC and in the Greek geographer Strabo's Geographika, compiled almost 2000 years ago. As cartography developed, geographers illuminated many aspects used today in the field; maps created by different European powers described the resources likely to be found in American, African, and Asian territories. The earliest travel journals included descriptions of the native peoples, the climate, the landscape, and the productivity of various locations. These early accounts encouraged the development of transcontinental trade patterns and ushered in the era of mercantilism.

The coffee trade is a worldwide industry.

World War II contributed to the popularization of geographical knowledge generally, and post-war economic recovery and development contributed to the growth of economic geography as a discipline. During environmental determinism's time of popularity, Ellsworth Huntington and his theory of climatic determinism, while later greatly criticized, notably influenced the field. Valuable contributions also came from location theorists such as Johann Heinrich von Thünen or Alfred Weber. Other influential theories include Walter Christaller's Central place theory, the theory of core and periphery.

Fred K. Schaefer's article critique of regionalism, made a large impact on the field: the article became a rallying point for the younger generation of economic geographers who were intent on reinventing the discipline as a science, and quantitative methods began to prevail in research. Well-known economic geographers of this period include William Garrison, Brian Berry, Waldo Tobler, Peter Haggett and William Bunge.

Contemporary economic geographers tend to specialize in areas such as location theory and spatial analysis (with the help of geographic information systems), market research, geography of transportation, real estate price evaluation, regional and global development, planning, Internet geography, innovation, social networks.

As economic geography is a very broad discipline, with economic geographers using many different methodologies in the study of economic phenomena in the world some distinct approaches to study have evolved over time:

- Theoretical economic geography focuses on building theories about spatial arrangement and distribution of economic activities.

- Regional economic geography examines the economic conditions of particular regions or countries of the world. It deals with economic regionalization as well as local economic development.

- Historical economic geography examines the history and development of spatial economic structure. Using historical data, it examines how centers of population and economic activity shift, what patterns of regional specialization and localization evolve over time and what factors explain these changes.

- Evolutionary economic geography adopts an evolutionary approach to economic geography. More specifically, Evolutionary Economic Geography uses concepts and ideas from evolutionary economics to understand the evolution of cities, regions, and other economic systems.

- Critical economic geography is an approach taken from the point of view of contemporary critical geography and its philosophy.

- Behavioral economic geography examines the cognitive processes underlying spatial reasoning, locational decision making, and behavior of firms and individuals.

Economic geography is sometimes approached as a branch of anthropogeography that focuses on regional systems of human economic activity. An alternative description of different approaches to the study of human economic activity can be organized around spatiotemporal analysis, analysis of production/consumption of economic items, and analysis of economic flow. Spatiotemporal systems of analysis include economic activities of region, mixed social spaces, and development.

Alternatively, analysis may focus on production, exchange, distribution, and consumption of items of economic activity. Allowing parameters of space-time and item to vary, a geographer may also examine material flow, commodity flow, population flow and information flow from different parts of the economic activity system. Through analysis of flow and production, industrial areas, rural and urban residential areas, transportation site, commercial service facilities and finance and other economic centers are linked together in an economic activity system.

Branches

Thematically, economic geography can be divided into these subdisciplines:

- Geography of agriculture.

  "It is traditionally considered the branch of economic geography that investigates those parts of the Earth's surface that are transformed by humans through primary sector activities. It thus focuses on structures of agricultural landscapes and asks for the processes that lead to these spatial patterns. While most research in this area concentrates rather on production than on consumption, a distinction can be made between nomothetic (e.g. distribution of spatial agricultural patterns and processes) and idiographic research (e.g. human-environment interaction and the shaping of agricultural landscapes). The latter approach of agricultural geography is often applied within regional geography."

- Geography of industry.

- Geography of international trade.

- Geography of resources.

- Geography of transport and communication.

- Geography of finance.

These areas of study may overlap with other geographical sciences.

Economists and Economic Geographers

Generally, spatially interested economists study the effects of space on the economy. Geographers, on the other hand, are interested in the economic processes' impact on spatial structures.

Moreover, economists and economic geographers differ in their methods in approaching spatial-economic problems in several ways. An economic geographer will often take a more holistic approach to the analysis of economic phenomena, which is to conceptualize a problem in terms of space, place, and scale as well as the overt economic problem that is being examined. The economist approach, according to some economic geographers, has the main drawback of homogenizing the economic world in ways economic geographers try to avoid.

New Economic Geography

With the rise of the New Economy, economic inequalities are increasing spatially. The New Economy, generally characterized by globalization, increasing use of information and communications technology, the growth of knowledge goods, and feminization, has enabled economic geographers to study social and spatial divisions caused by the rising New Economy, including the emerging digital divide.

The new economic geographies consist of primarily service-based sectors of the economy that use innovative technology, such as industries where people rely on computers and the internet. Within these is a switch from manufacturing-based economies to the digital economy. In these sectors, competition makes technological changes robust. These high technology sectors rely heavily on interpersonal relationships and trust, as developing things like software is very different from other kinds of industrial manufacturing—it requires intense levels of cooperation between many different people, as well as the use of tacit knowledge. As a result of cooperation becoming a necessity, there is a clustering in the high-tech new economy of many firms.

Social and Spatial Divisions

As characterized through the work of Diane Perrons, in Anglo-American literature, the

New Economy consists of two distinct types. New Economic Geography 1 (NEG1) is characterized by sophisticated spatial modelling. It seeks to explain uneven development and the emergence of industrial clusters. It does so through the exploration of linkages between centripetal and centrifugal forces, especially those of economies of scale.

New Economic Geography 2 (NEG2) also seeks to explain the apparently paradoxical emergence of industrial clusters in a contemporary context, however, it emphasizes relational, social, and contextual aspects of economic behaviour, particularly the importance of tacit knowledge. The main difference between these two types is NEG2's emphasis on aspects of economic behaviour that NEG1 considers intangible.

Both New Economic Geographies acknowledge transport costs, the importance of knowledge in a new economy, possible effects of externalities, and endogenous processes that generate increases in productivity. The two also share a focus on the firm as the most important unit and on growth rather than development of regions. As a result, the actual impact of clusters on a region is given far less attention, relative to the focus on clustering of related activities in a region.

However, the focus on the firm as the main entity of significance hinders the discussion of New Economic Geography. It limits the discussion in a national and global context and confines it to a smaller scale context. It also places limits on the nature of the firm's activities and their position within the global value chain. Further work done by Bjorn Asheim and Gernot Grabher challenges the idea of the firm through action-research approaches and mapping organizational forms and their linkages. In short, the focus on the firm in new economic geographies is undertheorized in NEG1 and undercontextualized in NEG2, which limits the discussion of its impact on spatial economic development.

Spatial divisions within these arising New Economic geographies are apparent in the form of the digital divide, as a result of regions attracting talented workers instead of developing skills at a local level. Despite increasing inter-connectivity through developing information communication technologies, the contemporary world is still defined through its widening social and spatial divisions, most of which are increasingly gendered. Danny Quah explains these spatial divisions through the characteristics of knowledge goods in the New Economy: goods defined by their infinite expansibility, weightlessness, and nonrivalry. Social divisions are expressed through new spatial segregation that illustrates spatial sorting by income, ethnicity, abilities, needs, and lifestyle preferences. Employment segregation is evidence by the overrepresentation of women and ethnic minorities in lower-paid service sector jobs. These divisions in the new economy are much more difficult to overcome as a result of few clear pathways of progression to higher-skilled work.

Geography of Finance

Geography of finance is a branch of economic geography that focuses on issues of financial globalization and the geographic patterns of finance. It studies the effects of state sovereignty, culture and different kinds of barriers that affect the spatial distribution of finance, such as uneven development and financial exclusion, and the global and local connectivity of financial flows and networks. Finally it also researches the creation of new financial centres around the world, both offshore and onshore.

Geography Matters

With the continuing process of globalization, some geographic barriers such as transportation costs of goods and capital are steadily decreasing. However, many other kinds of geographic distance are still very present and relevant to explain spatial differences. In geography of finance, researchers analyse the effects of this distance on the distribution of the financial system across the world. Fields of research include culture and education, technology, the effects of tacit knowledge and relational proximity and politics. An interesting issue in the latter is the increasing entanglement of banks and nations, which is closely related to geography of networks. Furthermore, researcher analyse how and how strongly that current spatial distribution of finance affects the allocation of funds, capital and credit across different regions.

Finance Matters

The relevance of economic geography is already quite established in the academic world and research on the topic is in full progress. However, geography of finance is now gaining individual focus, especially as the link between the financial economy and the real economy is losing strength. This is emphasized by the existence of economic bubbles, and the fact that the value of financial transactions is often multiple times larger than the real economy.

Recent Developments

The September 11 attacks that targeted the World Trade Centre buildings in New York City have drawn new attention to geography of finance. Even though cities have more often been damaged by natural disasters or terrorist attacks, this attack was so focused on the financial system and proved to have big effects on it. Therefore, the event caused people to rethink about the geographical organization of the financial services industry around the world, and academic attention on the importance of such densely organized financial districts. The financial crisis of 2007-2008 also caused interesting developments for the research in geography of finance. First of all it drew new attention to the field, as the crisis showed that local events could cause such a global financial crisis that affected even very small companies and (local) governments around the world. Secondly, the relocation of financial services that was already going on was amplified by this crisis, decreasing the importance of for example Wall Street to relatively new financial centres on the world.

HEALTH GEOGRAPHY

Health geography is the application of geographical information, perspectives, and methods to the study of health, disease, and health care.

The study of health geography has been influenced by repositioning medical geography within the field of social geography due to a shift towards a social model in health care, rather than a medical model. This advocates for the redefinition of health and health care away from prevention and treatment of illness only to one of promoting well-being in general. Under this model, some previous illnesses (e.g., mental ill health) are recognized as behavior disturbances only, and other types of medicine (e.g., complementary or alternative medicine and traditional medicine) are studied by the medicine researchers, sometimes with the aid of health geographers without medical education. This shift changes the definition of care, no longer limiting it to spaces such as hospitals or doctor's offices. Also, the social model gives priority to the intimate encounters performed at non-traditional spaces of medicine and healthcare as well as to the individuals as health consumers.

This alternative methodological approach means that medical geography is broadened to incorporate philosophies such as Marxian political economy, structuralism, social interactionism, humanism, feminism and queer theory.

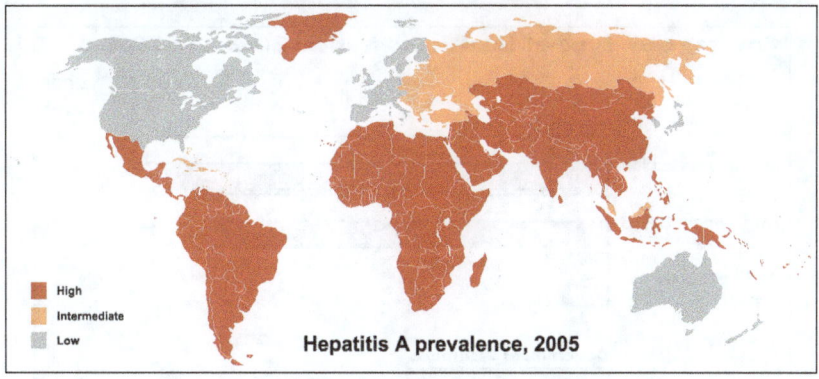

Hepatitis A prevalence worldwide.

The relationship between space and health dates back to Hippocrates, who stated that "airs, waters, places" all played significant roles impacting human health and history. A classic piece of research in health geography was done in 1854 as a cholera outbreak gripped a neighborhood in London. Death tolls rang around the clock and the people feared that they were being infected by vapors coming from the ground. John Snow predicted that if he could locate the source of the disease, it could be contained. He drew maps demonstrating the homes of people who had died of cholera and the locations of water pumps. He found that one pump, the public pump on Broad Street, was central to most of the victims. He concluded that infected water from the pump was the culprit. He instructed the authorities to remove the handle to the pump, making it unusable. As a result, the number of new cholera cases decreased.

Health geography is considered to be divided into two distinct elements. The first of which is focused on geographies of disease and ill health, involving descriptive research quantifying disease frequencies and distributions, and analytic research concerned with finding what characteristics make an individual or population susceptible to disease. This requires an understanding of epidemiology. The second component of health geography is the geography of health care, primarily facility location, accessibility, and utilization. This requires the use of spatial analysis and often borrows from behavioral economics.

Geographies of Disease and Ill Health

Health geographers are concerned with the prevalence of different diseases along a range of spatial scales from a local to global view, and inspects the natural world, in all of its complexity, for correlations between diseases and locations. This situates health geography alongside other geographical sub-disciplines that trace human-environment relations. Health geographers use modern spatial analysis tools to map the dispersion of various diseases, as individuals spread them amongst themselves, and across wider spaces as they migrate. Health geographers also consider all types of spaces as presenting health risks, from natural disasters, to interpersonal violence, stress, and other potential dangers.

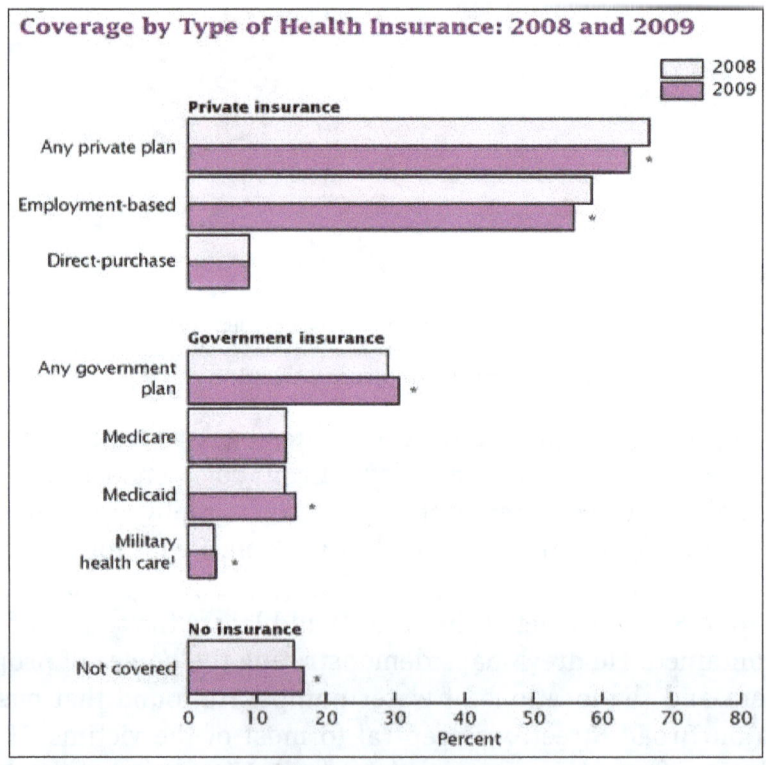

Health insurance coverage.

Geography of Health Care Provision

Although healthcare is a public good, it is not equally available to all individuals. Demand for public services is continuously increasing. People need advance knowledge and the latest prediction technology, that health geography offers. The latest example of such technology is Telemedicine. Many people in the United States are not able to access proper healthcare because of inequality in health insurance and the means to afford medical care.

Mobility and Disease Tracking: With the advent of mobile technology and its spread, it is now possible to track individual mobility. By correlating the movement of individuals through tracking the devices using access towers or other tracking systems, it is now possible to determine and even control disease spread. While privacy laws question the legality of tracking individuals, the commercial mobile service providers are using covert techniques or obtaining government waivers to allow permission to track people.

GEOGRAPHY OF FOOD

The geography of food is a field of human geography. It focuses on patterns of food production and consumption on the local to global scale. Tracing these complex patterns helps geographers understand the unequal relationships between developed and developing countries in relation to the innovation, production, transportation, retail and consumption of food. It is also a topic that is becoming increasingly charged in the public eye. The movement to reconnect the 'space' and 'place' in the food system is growing, spearheaded by the research of geographers.

Spatial variations in food production and consumption practices have been noted for thousands of years. In fact, Plato commented on the destructive nature of agriculture when he referred to the soil erosion from the mountainsides surrounding Athens, stating "Athens yielded far more abundant produce. In comparison of what then was, there are remaining only the bones of the wasted body; all the richer and softer parts of the soil having fallen away, and the mere skeleton of the land being left". Societies beyond those of ancient Greece have struggled under the pressure to feed expanding populations. The people of Easter Island, the Maya of Central America and most recently the inhabitants of Montana have been experiencing similar difficulties in production due to several interconnecting factors related to land and resource management. These events have been extensively studied by geographers and other interested parties (the study of food has not been confined to a single discipline, and has received attention from a huge range of diverse sources).

Modern geographers initially focused on food as an economic activity, especially in terms of agricultural geography. It was not until recently that geographers have turned their attention to food in a wider sense: "The emergence of an agro-food geography that seeks to examine issues along the food chain or within systems of food provision derives, in part, from the strengthening of political economy approaches in the 1980s".

Food Production

Food production was the first element of food to receive extensive attention from geographers in the field of cultural geography, particularly in agricultural geography.

Globally, the production of food is unequal. This is because there are two main components involved in sustenance production that are also distributed irregularly. These components are the environmental capacity of the area, and the human capacity. Environmental capacity is its ability 'to accommodate a particular activity or rate of an activity without unacceptable impact'. The climate, soil types, and availability of water affect it. Human capacity, in relation to food production, is the size of the population and the amount of agricultural skill within that population. When these two are at ideal levels and partnered with financial capital, the creation of intense agricultural infrastructure is possible, as the Green Revolution clearly portrays.

Simultaneously, the ability of a country to produce food is being severely impacted by a plethora of other factors:

1. Pests are becoming resistant to pesticides, or pesticides may be killing off the useful and necessary insects. Examples of this happening occur around the globe. Tanzania experienced a particularly horrible infection of armyworms in 2005. At the infections peak, there were over 1000 larva per square meter. In 2009, Liberia experienced a state of emergency when invading African armyworm caterpillars began what became a regional food crisis. The caterpillars traveled through 65 towns and 20 000 people were forced to leave their homes, markets, and farms. Losses like this can cost millions to billions, depending on size and duration, and have severe effects on food security. The FAO has created an international team, the Plant Production and Protection Division, which is attempting to 'reduce reliance on pesticides' and 'demonstrate that pesticide use often can be reduced considerably without affecting yields or farmer profits' in these, and other hard-struck areas.

Aeolian wind erosion.

2. Water stress, desertification, and erosion are leading to loss of arable land. Agricultural practices use the bulk of the Earth's fresh water – up to 70 percent – and those numbers are predicted to rise by 50-100 percent by 2025'. Countries are being forced to divert more water than ever before to irrigate their land. Hydroelectric dams and mega-canal projects are becoming the new standard for countries like Egypt that can no longer depend on rainfall or natural flood cycles. These water shortages are also causing a source of conflict between neighboring nations as they live with increasingly high levels of water scarcity. Policy responses to these events could be implemented in order to strengthen the socio-economic growth, human health statuses, and environmental sustainability of these areas. Combining current limitations with water and transitions from practices such as agroforestry and shifting cultivation makes land susceptible to aeolian erosion by weakening soil composition and exposing larger areas of land to destructive wind. Aeolian erosion largely effects deserted areas, reducing air quality, polluting water sources, and limiting fertility of nearby land.

3. Climate change is creating more extreme weather patterns, and agricultural practices are estimated to cause from 10 to 12 percent of greenhouses gas emissions. Warming will increase the rates of desertification and insect activity and agricultural zones near the equator may be lost. However, due to the uneven warming that will probably occur, higher latitudes are expected to warm up at faster rates than other areas of the globe. Scientists are now presenting the idea that areas in Canada and Siberia may become suitable for farming at the industrial scale, and that those areas will be able to account for any farmland that is lost at the equator. Conservative estimates place the shift of traditional crops (maize, grain, potatoes) northward at 50 to 70 kilometers a decade. It is also believed that non-traditional crops (berries, sunflowers, melons) could be established on the southern sides of these countries. Changes in climate may force humans to adapt, adopt new practices, and alter old habits to promote success in the uncertain age of climate change ahead.

Food Consumption

Criticisms of the industrialized food system regarding its inability to provide nutritious, ecologically sound, equitable food for the world's population has increased in recent history. Systems that are currently in place focus on providing relatively cheap food to millions, but often cost the Earth in terms of water and soil degradation, local food insecurity, animal welfare, rising obesity and health-related problems, and declining rural communities. Variations in diet and consumption practices on global and regional scales became the focus of geographers and economists with the vastly expanding population and widely publicized famines of the 1960s, and the food riots of 2007-2008 in 60 different countries. Due in part to these events, differences in the caloric intake of food and the composition of an average diet have been estimated and mapped for many countries since the 1960s.

Canada, USA, and Europe consume the highest amount of calories with an average per capita consumption of around 3400 calories daily. The recommended daily caloric intake for men and women living in these areas is 2500 and 2000 respectively. Studies focused on consumption patterns in these areas lay the blame for increased caloric intake on soft drink and fast food consumption, and decreased physical activity. Many developing countries are beginning to follow the leaders in rising caloric intake as they develop further due to increased availability of these high-impact items. Ballooning weight and associated health problems such as high blood pressure, high cholesterol, heart problems, and diabetes are being recorded in skyrocketing numbers.

Globally, consumption is still extremely uneven, with areas such as Sub-Saharan Africa still having some of the lowest rates of caloric intake per capita, often falling below the recommended levels. Much of this is due to lack of access of particular foods, which is a leading factor as to why much of the undernourished population is located in this region. In the world today, there are over 800 million people that are undernourished. The Democratic Republic of Congo holds the lowest average, at 1800 calories daily; however, averages do not represent the range of inequality between the best and worst fed people within a region. Currently, steps are being made to reduce caloric inequality. In parts of South Africa, the government has implemented a widespread electrification system featuring a free electricity allowance due to a study was conducted from 1991 to 2002 that found a positive increase in consumption habits within villages if given access to electricity. Access to electricity allowed for less time to be spent on menial tasks such as gathering firewood, and more time working on higher-level tasks that could increase income. In fact, villages often exceeded their electrical allowances.

TIME GEOGRAPHY

Time geography or time-space geography is an evolving transdisciplinary perspective on spatial and temporal processes and events such as social interaction, ecological interaction, social and environmental change, and biographies of individuals. Time geography "is not a subject area per se", but rather an integrative ontological framework and visual language in which space and time are basic dimensions of analysis of dynamic processes. Time geography was originally developed by human geographers, but today it is applied in multiple fields related to transportation, regional planning, geography, anthropology, time-use research, ecology, environmental science, and public health. According to Swedish geographer Bo Lenntorp: "It is a basic approach, and every researcher can connect it to theoretical considerations in her or his own way".

The Swedish geographer Torsten Hägerstrand created time geography in the mid-1960s based on ideas he had developed during his earlier empirical research on human migration patterns in Sweden. He sought "some way of finding out the workings of large socio-environmental mechanisms" using "a physical approach involving the study of how events occur in a time-space framework". Hägerstrand was inspired in part by conceptual advances in spacetime physics and by the philosophy of physicalism.

Hägerstrand's earliest formulation of time geography informally described its key ontological features: "In time-space the individual describes a path" within a situational context; "life paths become captured within a net of constraints, some of which are imposed by physiological and physical necessities and some imposed by private and common decisions". "It would be impossible to offer a comprehensive taxonomy of constraints seen as time-space phenomena", Hägerstrand said, but he "tentatively described" three important classes of constraints:

- Capability constraints: Limitations on the activity of individuals because of their biological structure and the tools they can command.

- Coupling constraints: Limitations that "define where, when, and for how long, the individual has to join other individuals, tools, and materials in order to produce, consume, and transact" (closely related to critical path analysis).

- Authority constraints: Limitations on the domain or "time-space entity within which things and events are under the control of a given individual or a given group".

Hägerstrand illustrated these concepts with novel forms of graphical notation (inspired in part by musical notation), such as:

- The space-time aquarium (or space-time cube), which displays individual paths in axonometric graphical projection of space and time coordinates.

- The space-time prism, which shows individuals' possible behavior in time-space given their capability constraints and coupling constraints.

- Bundles of paths, which are the conjunction of individual paths due in part to their capability constraints and coupling constraints, and which help to create "pockets of local order".

- Concentric tubes or rings of accessibility, which indicate certain capability constraints of a given individual, such as limited spatial size and limited manual, oral-auditive and visual range.

- Nested hierarchies of domains, which show the authority constraints for a given individual or a given group.

While this innovative visual language is an essential feature of time geography, Hägerstrand's colleague Bo Lenntorp emphasized that it is the product of an underlying ontology, and "not the other way around. The notation system is a very useful tool, but it is a rather poor reflection of a rich world-view. In many cases, the notational apparatus has been the hallmark of time geography. However, the underlying ontology is the most important feature". Time geography is not only about time-geographic diagrams, just as music is not only about musical notation. Hägerstrand later explained: "What is briefly alluded to here is a 4-dimensional world of forms. This cannot be completely graphically depicted. On the other hand one ought to be able to imagine it with sufficient clarity for it to be of guidance in empirical and theoretical research".

By 1981, geographers Nigel Thrift and Allan Pred were already defending time geography against those who would see it "merely as a rigid descriptive model of spatial and temporal organization which lends itself to accessibility constraint analysis (and related exercises in social engineering)". They argued that time geography is not just a model of constraints; it is a flexible and evolving way of thinking about reality that can complement a wide variety of theories and research methods. In the decades since then, Hägerstrand and others have made efforts to expand his original set of concepts. By the end of his life, Hägerstrand had ceased using the phrase "time geography" to refer to this way of thinking and instead used words like topoecology.

Later Developments in Time Geography

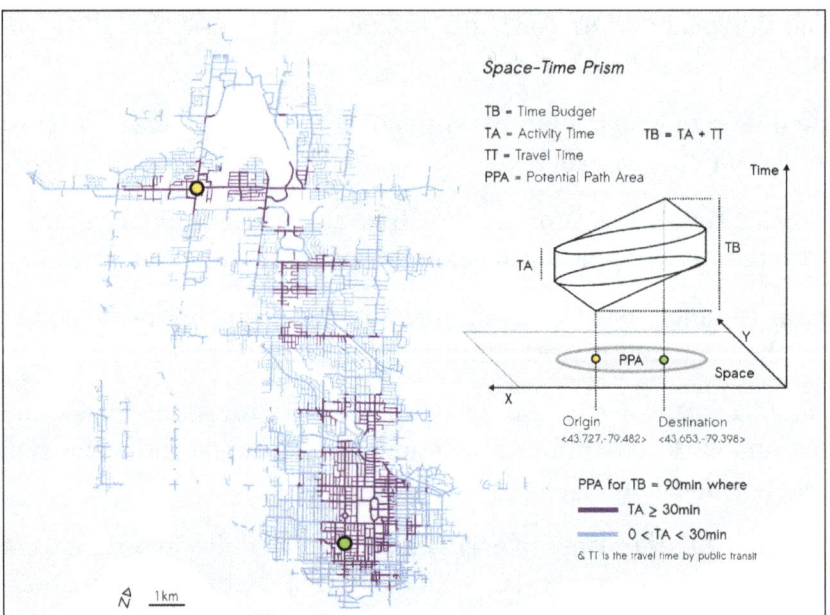

Schematic and example of a space-time prism using transit network data: On the right is a schematic diagram of a space-time prism, and on the left is a map of the potential path area for two different time budgets.

Since the 1980s, time geography has been used by researchers in the social sciences, the biological sciences, and in interdisciplinary fields.

In 1993, British geographer Gillian Rose noted that "time-geography shares the feminist interest in the quotidian paths traced by people, and again like feminism, links such paths, by thinking about constraints, to the larger structures of society". However, she noted that time geography had not been applied to issues important to feminists, and she called it a form of "social science masculinity". Over the following two decades, feminist geographers have revisited time geography and have begun to use it as a tool to address feminist issues.

GIS software has been developed to compute and analyze time-geographic problems at a variety of spatial scales. Such analyses have used different types of network datasets (such as walking networks, highway networks, and public transit schedules) as well as a variety of visualization strategies. Specialized software such as GeoTime has been developed to facilitate time-geographic visualization and visual analytics.

Time geography has also been used as a form of therapeutic assessment in mental health.

Benjamin Bach and colleagues have generalized the space-time cube into a framework for temporal data visualization that applies to all data that can be represented in two dimensions plus time.

References

- Human-geography-overview-1434505: thoughtco.com, Retrieved 20 January, 2020

- Whatmore, Sarah (2016). "Materialist returns: Practising cultural geography in and for a more-than-human world". Cultural Geographies. 13 (4): 600–609. doi:10.1191/1474474006cgj377oa

- Boschma, Ron; Frenken, Koen (2006). "Why is economic geography not an evolutionary science? Towards an evolutionary economic geography". Journal of Economic Geography. 6 (3): 273–302. doi:10.1093/jeg/lbi022

- Moseley, William (2014). An Introduction Human-Environment Geography. United Kingdom: Wiley Blackwell. p. 260. ISBN 9781405189316

- Munoz, J. Mark (2013). Handbook on the Geopolitics of Business. Edward Elgar Publishing : UK. ISBN 9780857939746

- Rogers, Alisdair; Castree, Noel; Kitchin, Rob (19 September 2013). "Population geography". A Dictionary of Human Geography. Oxford University Press. ISBN 9780199599868

- Tourism-geography: geographyrealm.com, Retrieved 27 August, 2020

3

Physical Geography

Physical geography is the study of processes and patterns in the natural environment. Biogeography, geomorphology, climatology, coastal geography, oceanography, palaeogeography, etc. are studied under its domain. This chapter has been carefully written to provide an easy understanding of these branches within physical geography.

Physical geography encompasses the geographic tradition known as the Earth sciences tradition. Physical geographers look at the landscapes, surface processes, and climate of the earth—all of the activity found in the four spheres (the atmosphere, hydrosphere, biosphere, and lithosphere) of our planet.

In contrast, cultural or human geography spends time studying why people locate where they do (including demographics) and how they adapt to and change the landscape in which they live. Someone studying cultural geography might also research how languages, religion, and other aspects of culture develop where people live; how those aspects are transmitted to others as people move; or how cultures change because of where they move.

The Four Spheres

The atmosphere itself has several layers to study, but the atmosphere as a topic under the lens of physical geography also includes research areas such as the ozone layer, the greenhouse effect, wind, jet streams, and weather.

The hydrosphere encompasses everything having to do with water, from the water cycle to acid rain, groundwater, runoff, currents, tides, and oceans.

The biosphere concerns living things on the planet and why they live where they do, with topics from ecosystems and biomes to food webs and the carbon and nitrogen cycles.

The study of the lithosphere includes geological processes, such as the formation of rocks, plate tectonics, earthquakes, volcanoes, soil, glaciers, and erosion.

Sub-branches of Physical Geography

Since the Earth and its systems are so complex, there are many sub-branches and even sub-sub-branches of physical geography as a research area, depending on how granularly the categories are divided. They also have overlap between them or with other disciplines, such as geology.

Geographical researchers will never be at a loss of something to study, as they often need to understand multiple areas to inform their own targeted research.

- Geomorphology: The study of Earth's landforms and its surface's processes—and how these processes change and have changed Earth's surface—such as erosion, landslides, volcanic activity, earthquakes, and floods.

- Hydrology: The study of the water cycle, including water distribution across the planet in lakes, rivers, aquifers, and groundwater; water quality; drought effects; and the probability of flooding in a region. Potamology is the study of rivers.

- Glaciology: The study of glaciers and ice sheets, including their formation, cycles, and effect on Earth's climate.

- Biogeography: The study of the distribution of life forms across the planet, relating to their environments; this field of study is related to ecology, but it also looks into the past distribution of life forms as well, as found in the fossil record.

- Meteorology: The study of Earth's weather, such as fronts, precipitation, wind, storms, and the like, as well as forecasting short-term weather based on available data.

- Climatology: The study of Earth's atmosphere and climate, how it has changed over time, and how humans have affected it.

- Pedology: The study of soil, including types, formation, and regional distribution over Earth.

- Paleogeography: The study of historical geographies, such as the location of the continents over time, through looking at geological evidence, such as the fossil record.

- Coastal geography: The study of the coasts, specifically concerning what happens where land and water meet.

- Oceanography: The study of the world's oceans and seas, including aspects such as floor depths, tides, coral reefs, underwater eruptions, and currents. Exploration and mapping is a part of oceanography, as is research into the effects of water pollution.

- Quaternary science: The study of the previous 2.6 million years on Earth, such as the most recent ice age and Holocene period, including what it can tell us about the change in Earth's environment and climate.

- Landscape ecology: The study of how ecosystems interact with and affect each other in an area, especially looking at the effects of the uneven distribution of landforms and species in these ecosystems (spatial heterogeneity).

- Geomatics: The field that gathers and analyzes geographic data, including the gravitational force of Earth, the motion of the poles and Earth's crust, and ocean tides (geodesy). In geomatics, researchers use the Geographic Information System (GIS), which is a computerized system for working with map-based data.

- Environmental geography: The study of the interactions between people and their environment and the resulting effects, both on the environment and on the people; this field bridges physical geography and human geography.

- Astronomical geography or astronography: The study of how the sun and moon affect the Earth as well as our planet's relationship to other celestial bodies.

Knowing about the physical geography of Earth is important for every serious student studying the planet because the natural processes of Earth affect the distribution of resources (from carbon dioxide in the air to freshwater on the surface to minerals deep underground) and the conditions of the human settlement. Anyone studying processes involving Earth and its processes is working within the confines of its physical geography. These natural processes have resulted in a plethora of varied effects on human populations throughout the millennia.

IMPORTANCE OF PHYSICAL GEOGRAPHY

Physical Geography plays a very vital role in the very existence of human beings. Every student who studies about the planet earth should also study Physical Geography. This is because physical geography involves the study of the natural processes of the earth.

- In Physical Geography, you will study all about the various elements of nature.

- By study in this area, you can learn about the weather and how it's changing (and the possible results of those transforms) have an effect on people now and can help plan for the prospect.

- Geography doesn't just find out whether humans can live in a definite area or not, it also determines people's lifestyles, as they adjust to the obtainable food and atmosphere patterns.

- It also deals with the interaction of our planet Earth with that of the sun, moon, stars, space.

- It also covers all aspects of the earth in relation to the climatic conditions, the seasons, atmospheric force and composition, the effects of the wind, storms, rain, snow, and other aspects.

- It deals with the various climatic zones, flora and fauna, wild animals, the hydrological cycles, precipitation, the wind pressures, the microclimates, soil erosion, desserts, etc.

The study of physical Geography is essential for the sufficient allocation of the natural resources on the earth. It is essential for enabling human resolution as per the adjacent conditions and to be improved informed in order to protect our planet earth.

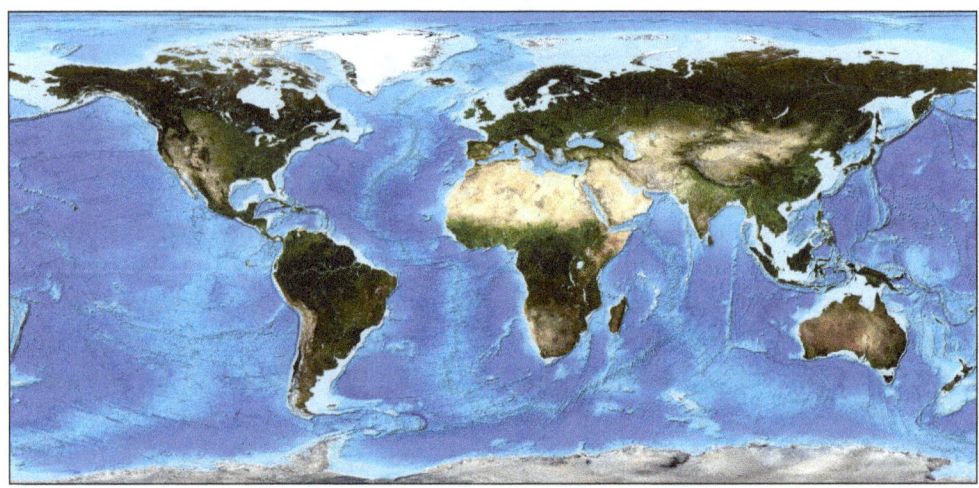

Time provides maturity to soils and facilitates in the maturity of soil shapes. Each component is significant for human beings. Landforms give the base on which human activities are placed. The plains are utilized for agriculture. Plateaus provide forests and minerals. Mountains provide pastures, forests, tourist spots and are sources of rivers providing water to lowlands. Climate influences our house types, clothing and food habits. The weather has a thoughtful consequence on vegetation, cropping pattern, domestic animals farming, and several industries, etc. Human beings have developed technologies which adjust climatic elements in a restricted space such as air conditioners and coolers. Temperature and rainfall make sure the compactness of forests and quality of grassland.

For, example, monsoonal rainfall sets the cultivation measure in motion. Rainfall recharges the soil water aquifer which later provides water for agriculture and household use. We study oceans which are the storehouse of resources. Besides fish and other sea-food, oceans are rich in mineral resources. It has developed the technology for collecting manganese nodules from the oceanic bed. Soils are renewable resources, which influence a number of economic activities such as agriculture. The fertility of the soil is both naturally determined and culturally induced. Soils also make available the foundation for the biosphere accommodating plants, animals, and microorganisms.

So, Geography is the study of the Earth's overall surroundings and the creature that live in those environments. Physical geographers concern geomorphologic principals to study how landforms have changed in history, but ever more such principals are significant for contemporary applications.

MAJOR LANDFORMS

Plateau

In geology and physical geography, a plateau also called a high plain or a tableland, is an area of a highland, usually consisting of relatively flat terrain, that is raised significantly above the surrounding area, often with one or more sides with deep hills. Plateaus can be formed by a number of processes, including upwelling of volcanic magma, extrusion of lava, and erosion by water and glaciers. Plateaus are classified according to their surrounding environment as intermontane, piedmont, or continental. A few plateaus may have a small flat top while others have wide ones.

Formation

Plateaus can be formed by a number of processes, including upwelling of volcanic magma, extrusion of lava, and erosion by water and glaciers.

Volcanic

The Pajarito Plateau is an example of a volcanic plateau.

Volcanic plateaus are produced by volcanic activity. The Columbia Plateau in the northwestern United States is an example. They may be formed by upwelling of volcanic magma or extrusion of lava.

The underlining mechanism in forming plateaus from upwelling starts when magma rises from the mantle, causing the ground to swell upward. In this way, large, flat areas of rock are uplifted to form a plateau. For plateaus formed by extrusion, the rock is built up from lava spreading outward from cracks and weak areas in the crust.

Erosion

Plateaus can also be formed by the erosional processes of glaciers on mountain ranges, leaving them sitting between the mountain ranges. Water can also erode mountains and other landforms down into plateaus. Dissected plateaus are highly eroded plateaus cut by rivers and broken by deep narrow valleys. Computer modeling studies suggest that high plateaus may also be partially a result from the feedback between tectonic deformation and dry climatic conditions created at the lee side of growing orogens.

Classification

Plateaus are classified according to their surrounding environment:

- Intermontane plateaus are the highest in the world, bordered by mountains. The Tibetan Plateau is one such plateau.

- Lava or volcanic plateaus are the plateau that occur in areas of widespread volcanic eruptions. The magma that comes out through narrow cracks or fissures in the crust spread over large area and solidifies. These layers of lava sheets form lava or volcanic plateaus. The Antrim plateau in Northern Ireland, The Deccan Plateau in India and the Columbia Plateau in the United States are examples of lava plateaus.

- Piedmont plateaus are bordered on one side by mountains and on the other by a plain or a sea. The Piedmont Plateau of the Eastern United States between the Appalachian Mountains and the Atlantic Coastal Plain is an example.

- Continental plateaus are bordered on all sides by plains or oceans, forming away from the mountains. An example of a continental plateau is the Antarctic Plateau in East Antarctica.

Large Plateaus

The largest and highest plateau in the world is the Tibetan Plateau, sometimes metaphorically described as the "Roof of the World", which is still being formed by the collisions of the Indo-Australian and Eurasian tectonic plates. The Tibetan plateau covers approximately 2,500,000 km^2 (970,000 sq mi), at about 5,000 m (16,000 ft) above sea level. The plateau is sufficiently high to reverse the Hadley cell convection cycles and to drive the monsoons of India towards the south.

The second-highest plateau is the Deosai Plateau of the Deosai National Park (also known as Deoasai Plains) at an average elevation of 4,114 m (13,497 ft). It is located in the Astore and Skardu districts of Gilgit-Baltistan, in northern Pakistan. Deosai means 'the land of giants'. The park protects an area of 3,000 km^2 (1,200 sq mi). It is known for its rich flora and fauna of the Karakoram-West Tibetan Plateau alpine steppe ecoregion. In spring it is covered by sweeps of wildflowers and a wide variety of butterflies. The highest point in Deosai is Deosai Lake, or Sheosar Lake from the Shina language meaning "Blind lake" (Sheo – Blind, Sar – lake) near the Chilim Valley. The lake lies at an elevation of 4,142 m (13,589 ft), one of the highest lakes in the world, and is 2.3 km (1.4 mi) long, 1.8 km (1.1 mi) wide, and 40 m (130 ft) deep on average.

Hardangervidda, the largest plateau.

Some other major plateaus in Asia are: Najd in the Arabian Peninsula elevation 762 to 1,525 m (2,500 to 5,003 ft), Armenian Highlands ($\approx$400,000 km² (150,000 sq mi), elevation 900–2,100 metres (3,000–6,900 ft)), Iranian plateau ($\approx$3,700,000 km² (1,400,000 sq mi), elevation 300–1,500 metres (980–4,920 ft)), Anatolian Plateau, Mongolian Plateau ($\approx$2,600,000 km² (1,000,000 sq mi), elevation 1,000–1,500 metres (3,300–4,900 ft)), and the Deccan Plateau ($\approx$1,900,000 km² (730,000 sq mi), elevation 300–600 metres (980–1,970 ft)).

Another very large plateau is the icy Antarctic Plateau, which is sometimes referred to as the Polar Plateau, home to the geographic South Pole and the Amundsen-Scott South Pole Station, which covers most of East Antarctica where there are no known mountains but rather 3,000 m (9,800 ft) high of superficial ice and which spreads very slowly toward the surrounding coastline through enormous glaciers. This polar ice cap is so massive that the echolocation sound measurements of ice thickness have shown that large parts of the Antarctic "dry land" surface have been pressed below sea level. Thus, if that same ice cap were suddenly removed, the large areas of the frozen white continent would be flooded by the surrounding Antarctic Ocean or Southern Ocean. On the other hand, were the ice cap melts away too gradually, the surface of the land beneath it would gradually rebound away through isostasy from the center of the Earth and that same land would ultimately rise above sea level.

A large plateau in North America is the Colorado Plateau, which covers about 337,000 km² (130,000 sq mi) in Colorado, Utah, Arizona and New Mexico.

Bogotá, is located in a high plateau, over 8,600 ft (2,600 m) high.

In northern Arizona and southern Utah the Colorado Plateau is bisected by the Colorado River and the Grand Canyon. How this came to be is that over 10 million years ago, a river was already there, though not necessarily on exactly the same course. Then, subterranean geological forces caused the land in that part of North America to gradually rise by about a centimeter per year for millions of years. An unusual balance occurred: the river that would become the Colorado River was able to erode into the crust of the Earth at a nearly equal rate to the uplift of the plateau. Now, millions of years later, the North Rim of the Grand Canyon is at an elevation of about 2,450 m (8,040 ft) above sea level, and the South Rim of the Grand Canyon is about 2,150 m (7,050 ft) above sea level. At its deepest, the Colorado River is about 1,830 m (6,000 ft) below the level of the North Rim.

Another high altitude plateau in North America is the Mexican Plateau. With an area of 601,882 km² (232,388 sq mi) and average height of 1,825 m, it is the home of more than 70 million people.

A tepui, or tepuy is a table-top mountain or mesa found in the Guiana Highlands of South America, especially in Venezuela and western Guyana. The word tepui means "house of the gods" in the native tongue of the Pemon, the indigenous people who inhabit the Gran Sabana.

Tepuis can be considered minute plateaus and tend to be found as isolated entities rather than in connected ranges, which makes them the host of a unique array of endemic plant and animal species. Some of the most outstanding tepuis are Neblina, Autana, Auyan and Mount Roraima. They are typically composed of sheer blocks of Precambrian quartz arenite sandstone that rise abruptly from the jungle, giving rise to spectacular natural scenery. Auyantepui is the source of Angel Falls, the world's tallest waterfall.

The Colombian capital city of Bogota sits on an Andean plateau known as the Altiplano Cundiboyacense roughly the size of Switzerland. Averaging a height of 2,600 m (8,500 ft) above sea level, this northern Andean plateau is situated in the country's eastern range and is divided into three main flat regions: the Bogotá savanna, the valleys of Ubaté and Chiquinquirá, and the valleys of Duitama and Sogamoso.

The parallel Sierra of Andes delimit one of the world highest plateaux: the Altiplano, (Spanish for "high plain"), Andean Plateau or Bolivian Plateau. It lies in west-central South America, where the Andes are at their widest, is the most extensive area of high plateau on Earth outside of Tibet. The bulk of the Altiplano lies within Bolivian and Peruvian territory while its southern parts lie in Chile. The Altiplano plateau hosts several cities like Puno, Oruro, El Alto and La Paz the administrative seat of Bolivia. Northeastern Altiplano is more humid than the Southwestern, the latter of which hosts several salares, or salt flats, due to its aridity. At the Bolivia-Peru border lies Lake Titicaca, the largest lake in South America.

The highest African plateau is the Ethiopian Highlands which cover the central part of Ethiopia. It forms the largest continuous area of its altitude in the continent, with little of its surface falling below 1500 m (4,921 ft), while the summits reach heights of up to 4550 m (14,928 ft). It is sometimes called the Roof of Africa due to its height and large area.

Another example is the Highveld which is the portion of the South African inland plateau which has an altitude above approximately 1500 m, but below 2100 m, thus excluding the Lesotho mountain regions. It is home to some of largest South-African urban agglomerations.

In Egypt are the Giza Plateau and Galala Mountain, which was once called Gallayat Plateaus, raising 3,300 above sea level.

The Western Plateau, part of the Australian Shield, is an ancient craton covering much of the continent's southwest, an area of some 700,000 square kilometres. It has an average elevation of between 305 and 460 m.

The North Island Volcanic Plateau is an area of high land occupying much of the centre of the North Island of New Zealand, with volcanoes, lava plateaus, and crater lakes, the most notable of which is the country's largest lake, Lake Taupo. The plateau stretches approximately 100 km east to west and 130 km north to south. The majority of the plateau is more than 600 m above sea level.

Plain

A plain is a broad area of relatively flat land. Plains are one of the major landforms, or types of land, on Earth. They cover more than one-third of the world's land area. Plains exist on every continent.

Grasslands

Many plains, such as the Great Plains that stretch across much of central North America, are grasslands. A grassland is a region where grass is the main type of vegetation.

In North America, temperate grasslands—those in places with warm summers and cold winters—are often called prairies. In areas with little rain and snow, short grasses grow. In areas that receive more rain and snow, tall grasses can grow 1.5 meters (5 feet) high. However, most tallgrass prairies have been plowed under and are now farmland or pasture.

The Great Plains have supported a wide variety of cultures for thousands of years. The so-called "Plains Indians" are actually more than two dozen tribes. Communities include Blackfoot, native to the Canadian province of Alberta; Arapaho, whose center today is in the U.S. state of Wyoming; and Kickapoo, many of whom live today in the Mexican state of Coahuila.

In Asia and eastern Europe, temperate grasslands are called steppes. Steppes usually do not receive enough rain for tall grasses and trees to grow.

Tropical grasslands are called savannas. Savannas exist in places that are warm throughout the year. They often have scattered trees. Savannas such as the Serengeti plains stretch across much of central Africa. They are also found in Australia, South America, and southern North America.

Not all plains are grasslands. Some, such as Mexico's Tabasco Plain, are forested. Forested plains have different types of trees, shrubs, and other vegetation.

Deserts can also be plains. Parts of the Sahara, a great desert in North Africa, are plains.

In the Arctic, where the ground is frozen, plains are called tundra. Despite the cold, many plants survive here, including shrubs and moss.

Plain Formation

Plains form in many different ways. Some plains form as ice and water erodes, or wears away, the dirt and rock on higher land. Water and ice carry the bits of dirt, rock, and other material, called sediment, down hillsides to be deposited elsewhere. As layer upon layer of this sediment is laid down, plains form.

Volcanic activity can also form plains. Lava plains form when lava pushes up from below ground and flows across the land. The earth in a lava plain is often much darker than the surrounding soil. The dark earth is a result of the lava, mostly a dark-colored mineral called basalt, broken down into tiny particles over millions of years.

The movement of rivers sometimes forms plains. Many rivers run through valleys. As rivers move from side to side, they gradually erode the valley, creating broad plains.

As a river floods, it overflows its bank. The flood carries mud, sand, and other sediment out over the land. After the water withdraws, the sediment remains. If a river floods repeatedly, over time this sediment will build up into a flood plain. Flood plains are often rich in nutrients and create fertile farmland. The flood plain surrounding Africa's Nile River has helped Egyptian civilization thrive for thousands of years.

Alluvial plains form at the base of mountains. Water carrying sediment flows downhill until it hits flat land. There, it spreads out, depositing the sediment in the shape of a fan. The Huang He River in China has created an alluvial plain that covers about 409,500 square kilometers (158,000 square miles). Because much of the sediment the Huang He carries is yellowish in color, it is also called the Yellow River.

Many rivers deposit their sediment in the ocean. As the sediment builds up, it might eventually rise above sea level, forming a coastal plain. The Atlantic Coastal Plain stretches along much of the eastern coast of North America. These broad underwater plains slope gently down beneath the water.

Abyssal plains are found at the bottom of the ocean. These plains are 5,000 to 7,000 meters (16,400 to 23,000 feet) below sea level, so scientists have a hard time studying them. But scientists say abyssal plains are among the flattest, smoothest places on Earth.

Mountain

Mount Ararat.

A mountain is a large landform that rises above the surrounding land in a limited area, usually in the form of a peak. A mountain is generally steeper than a hill. Mountains are formed through tectonic forces or volcanism. These forces can locally raise the surface of the earth. Mountains erode slowly through the action of rivers, weather conditions, and glaciers. A few mountains are isolated summits, but most occur in huge mountain ranges.

High elevations on mountains produce colder climates than at sea level. These colder climates strongly affect the ecosystems of mountains: different elevations have different plants and animals. Because of the less hospitable terrain and climate, mountains tend to be used less for agriculture and more for resource extraction and recreation, such as mountain climbing and skiing.

The highest mountain on Earth is Mount Everest in the Himalayas of Asia, whose summit is 8,850 m (29,035 ft) above mean sea level. The highest known mountain on any planet in the Solar System is Olympus Mons on Mars at 21,171 m (69,459 ft).

There is no universally accepted definition of a mountain. Elevation, volume, relief, steepness, spacing and continuity have been used as criteria for defining a mountain. A mountain is defined as "a natural elevation of the earth surface rising more or less abruptly from the surrounding level and attaining an altitude which, relatively to the adjacent elevation, is impressive or notable".

Peaks of Mount Kenya.

Whether a landform is called a mountain may depend on local usage. Mount Scott outside Lawton, Oklahoma, USA, is only 251 m (823 ft) from its base to its highest point. Whittow states "Some authorities regard eminences above 600 metres (2,000 ft) as mountains, those below being referred to as hills".

In the United Kingdom and the Republic of Ireland, a mountain is usually defined as any summit at least 2,000 feet (610 m) high, which accords with the official UK government's definition that a mountain, for the purposes of access, is a summit of 2,000 feet (610 m) or higher. In addition, some definitions also include a topographical prominence requirement, typically 100 or 500 feet (30 or 152 m). At one time the U.S. Board on Geographic Names defined a mountain as being 1,000 feet (300 m) or taller, but has abandoned the definition since the 1970s. Any similar landform lower than this height was considered a hill. However, today, the United States Geological Survey (USGS) concludes that these terms do not have technical definitions in the US.

Mount Wilhelm.

The UN Environmental Programme's definition of "mountainous environment" includes any of the following:

- Elevation of at least 2,500 m (8,200 ft);

- Elevation of at least 1,500 m (4,900 ft), with a slope greater than 2 degrees;

- Elevation of at least 1,000 m (3,300 ft), with a slope greater than 5 degrees;

- Elevation of at least 300 m (980 ft), with a 300 m (980 ft) elevation range within 7 km (4.3 mi).

Using these definitions, mountains cover 33% of Eurasia, 19% of South America, 24% of North America, and 14% of Africa. As a whole, 24% of the Earth's land mass is mountainous.

Geology

There are three main types of mountains: volcanic, fold, and block. All three types are formed from plate tectonics: when portions of the Earth's crust move, crumple, and dive. Compressional forces, isostatic uplift and intrusion of igneous matter forces surface rock upward, creating a landform higher than the surrounding features. The height of the feature makes it either a hill or, if higher and steeper, a mountain. Major mountains tend to occur in long linear arcs, indicating tectonic plate boundaries and activity.

Volcanoes

Volcanoes are formed when a plate is pushed below another plate, or at a mid-ocean ridge or hotspot. At a depth of around 100 km, melting occurs in rock above the slab (due to the addition of water), and forms magma that reaches the surface. When the magma reaches the surface, it often builds a volcanic mountain, such as a shield volcano or a stratovolcano. Examples of volcanoes include Mount Fuji in Japan and Mount Pinatubo in the Philippines. The magma does not have to reach the surface in order to create a mountain: magma that solidifies below ground can still form dome mountains, such as Navajo Mountain in the US.

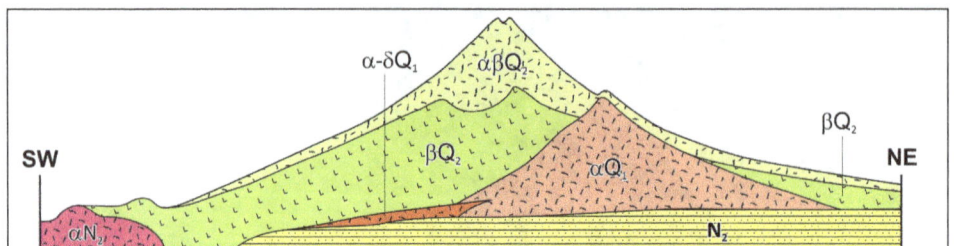

Geological cross-section of Fuji volcano.

Fold Mountains

Illustration of mountains that developed on a fold that thrusted.

Fold mountains occur when two plates collide: shortening occurs along thrust faults and the crust is overthickened. Since the less dense continental crust "floats" on the denser mantle rocks beneath, the weight of any crustal material forced upward to form hills, plateaus or mountains must be balanced by the buoyancy force of a much greater volume forced downward into the mantle. Thus the continental crust is normally much thicker under mountains, compared to lower lying areas. Rock can fold either symmetrically or asymmetrically. The upfolds are anticlines and the downfolds are synclines: in asymmetric folding there may also be recumbent and overturned folds. The Balkan Mountains and the Jura Mountains are examples of fold mountains.

Block Mountains

Pirin Mountain, Bulgaria, part of the fault-block Rila-Rhodope massif.

Block mountains are caused by faults in the crust: a plane where rocks have moved past each other. When rocks on one side of a fault rise relative to the other, it can form a mountain. The uplifted blocks are block mountains or horsts. The intervening dropped blocks are termed graben: these can be small or form extensive rift valley systems. This form of landscape can be seen in East Africa, the Vosges, the Basin and Range Province of Western North America and the Rhine valley. These areas often occur when the regional stress is extensional and the crust is thinned.

Erosion

The Catskills represent an eroded plateau.

During and following uplift, mountains are subjected to the agents of erosion (water, wind, ice, and gravity) which gradually wear the uplifted area down. Erosion causes the surface of mountains to be younger than the rocks that form the mountains themselves. Glacial processes produce characteristic landforms, such as pyramidal peaks, knife-edge arêtes, and bowl-shaped cirques that can contain lakes. Plateau mountains, such as the Catskills, are formed from the erosion of an uplifted plateau.

In Earth science, erosion is the action of surface processes (such as water flow or wind) that removes soil, rock, or dissolved material from one location on the Earth's crust, and then transport it away to another location (not to be confused with weathering which involves no movement). The particulate breakdown of rock or soil into clastic sediment is referred to as physical or mechanical erosion; this contrasts with chemical erosion, where soil or rock material is removed from an area by its dissolving into a solvent (typically water), followed by the flow away of that solution. Eroded sediment or solutes may be transported just a few millimetres, or for thousands of kilometres.

Climate

Climate in the mountains becomes colder at high elevations, due to an interaction between radiation and convection. Sunlight in the visible spectrum hits the ground and heats it. The ground then heats the air at the surface. If radiation were the only way to transfer heat from the ground to space, the greenhouse effect of gases in the atmosphere would keep the ground at roughly 333 K (60 °C; 140 °F), and the temperature would decay exponentially with height.

A combination of high latitude and high altitude makes the northern
Urals in picture to have climatic conditions that make the ground barren.

However, when air is hot, it tends to expand, which lowers its density. Thus, hot air tends to rise and transfer heat upward. This is the process of convection. Convection comes to equilibrium when a parcel of air at a given altitude has the same density as its surroundings. Air is a poor conductor of heat, so a parcel of air will rise and fall without exchanging heat. This is known as an adiabatic process, which has a characteristic pressure-temperature dependence. As the pressure gets lower, the temperature decreases. The rate of decrease of temperature with elevation is known as the adiabatic lapse rate, which is approximately 9.8 °C per kilometre (or 5.4 °F (−14.8 °C) per 1000 feet) of altitude.

Note that the presence of water in the atmosphere complicates the process of convection. Water vapor contains latent heat of vaporization. As air rises and cools, it eventually becomes saturated and cannot hold its quantity of water vapor. The water vapor condenses (forming clouds), and releases heat, which changes the lapse rate from the dry adiabatic lapse rate to the moist adiabatic lapse rate (5.5 °C per kilometre or 3 °F (−16 °C) per 1000 feet) The actual lapse rate can vary by altitude and by location.

Therefore, moving up 100 metres on a mountain is roughly equivalent to moving 80 kilometres (45 miles or 0.75° of latitude) towards the nearest pole. This relationship is only approximate, however, since local factors such as proximity to oceans (such as the Arctic Ocean) can drastically modify the climate. As the altitude increases, the main form of precipitation becomes snow and the winds increase.

The effect of the climate on the ecology at an elevation can be largely captured through a combination of amount of precipitation, and the biotemperature, as described by Leslie Holdridge in 1947. Biotemperature is the mean temperature; all temperatures below 0 °C (32 °F) are considered to be 0 °C. When the temperature is below 0 °C, plants are dormant, so the exact temperature is unimportant. The peaks of mountains with permanent snow can have a biotemperature below 1.5 °C (34.7 °F).

Mountains retain snow much longer than lower elevations.

Ecology

An alpine mire.

The colder climate on mountains affects the plants and animals residing on mountains. A particular set of plants and animals tend to be adapted to a relatively narrow range of climate. Thus, ecosystems tend to lie along elevation bands of roughly constant climate. This is called altitudinal zonation. In regions with dry climates, the tendency of mountains to have higher precipitation as well as lower temperatures also provides for varying conditions, which enhances zonation.

Some plants and animals found in altitudinal zones tend to become isolated since the conditions above and below a particular zone will be inhospitable and thus constrain their movements or dispersal. These isolated ecological systems are known as sky islands.

Altitudinal zones tend to follow a typical pattern. At the highest elevations, trees cannot grow, and whatever life may be present will be of the alpine type, resembling tundra. Just below the tree line, one may find subalpine forests of needleleaf trees, which can withstand cold, dry conditions. Below that, montane forests grow. In the temperate portions of the earth, those forests tend to be needleleaf trees, while in the tropics, they can be broadleaf trees growing in a rain forest.

Mountains and Humans

The highest known permanently tolerable altitude is at 5,950 metres (19,520 ft). At very high altitudes, the decreasing atmospheric pressure means that less oxygen is available for breathing, and there is less protection against solar radiation (UV). Above 8,000 metres (26,000 ft) elevation, there is not enough oxygen to support human life. This is known as the "death zone". The summits of Mount Everest and K2 are in the death zone.

Mountain Societies and Economies

Mountains are generally less preferable for human habitation than lowlands, because of harsh weather and little level ground suitable for agriculture. While 7% of the land area of Earth is above 2,500 metres (8,200 ft), only 140 million people live above that altitude and only 20-30 million people above 3,000 metres (9,800 ft) elevation. About half of mountain dwellers live in the Andes, Central Asia, and Africa.

With limited access to infrastructure, only a handful of human communities exist above 4,000 metres (13,000 ft) of elevation. Many are small and have heavily specialized economies, often relying on industries such as agriculture, mining, and tourism. An example of such a specialized town is La Rinconada, Peru, a gold-mining town and the highest elevation human habitation at 5,100 metres (16,700 ft). A counterexample is El Alto, Bolivia, at 4,150 metres (13,620 ft), which has a highly diverse service and manufacturing economy and a population of nearly 1 million.

Traditional mountain societies rely on agriculture, with higher risk of crop failure than at lower elevations. Minerals often occur in mountains, with mining being an important component of the economics of some montane societies. More recently, tourism supports mountain communities, with some intensive development around attractions such as national parks or ski resorts. About 80% of mountain people live below the poverty line.

Most of the world's rivers are fed from mountain sources, with snow acting as a storage mechanism for downstream users. More than half of humanity depends on mountains for water.

In geopolitics mountains are often seen as preferable "natural boundaries" between polities.

Mountaineering

Mountaineering, mountain climbing, or alpinism is the sport, hobby or profession of hiking, skiing, and climbing mountains. While mountaineering began as attempts to reach the highest point of unclimbed big mountains it has branched into specializations that address different aspects of the mountain and consists of three areas: rock-craft, snow-craft and skiing, depending on whether the route chosen is over rock, snow or ice. All require experience, athletic ability, and technical knowledge to maintain safety.

Mountain climbers ascending Mount Rainier.

Mountains as Sacred Places

Mountains often play a significant role in religions and philosophical beliefs. There are for example a number of sacred mountains within Greece such as Mount Olympos which was held to be the home of the gods. In Japanese culture, the 3,776.24 m (12,389 ft) volcano of Mount Fuji is also held to be sacred with tens of thousands of Japanese ascending it each year. In Ireland, pilgrimages are made up the 952 metres (3,123 ft) Mount Brandon by Irish Catholics. The Himalayan peak of Nanda Devi is associated with the Hindu goddesses Nanda and Sunanda; it has been off-limits to climbers since 1983.

Superlatives

Mount Everest, the highest peak on Earth.

Heights of mountains are typically measured above sea level. Using this metric, Mount Everest is the highest mountain on Earth, at 8,848 metres (29,029 ft). There are at least 100 mountains with heights of over 7,200 metres (23,622 ft) above sea level, all of which are located in central and southern Asia. The highest mountains above sea level are generally not the highest above the surrounding terrain. There is no precise definition of surrounding base, but Denali, Mount Kilimanjaro and Nanga Parbat are possible candidates for the tallest mountain on land by this measure. The bases of mountain islands are below sea level, and given this consideration Mauna Kea (4,207 m (13,802 ft) above sea level) is the world's tallest mountain and volcano, rising about 10,203 m (33,474 ft) from the Pacific Ocean floor.

The highest mountains are not generally the most voluminous. Mauna Loa (4,169 m or 13,678 ft) is the largest mountain on Earth in terms of base area (about 2,000 sq mi or 5,200 km²) and volume (about 18,000 cu mi or 75,000 km³). Mount Kilimanjaro is the largest non-shield volcano in terms of both base area (245 sq mi or 635 km²) and volume (1,150 cu mi or 4,793 km³). Mount Logan is the largest non-volcanic mountain in base area (120 sq mi or 311 km²).

The highest mountains above sea level are also not those with peaks farthest from the centre of the Earth, because the figure of the Earth is not spherical. Sea level closer to the equator is several miles farther from the centre of the Earth. The summit of Chimborazo, Ecuador's tallest mountain, is usually considered to be the farthest point from the Earth's centre, although the southern summit of Peru's tallest mountain, Huascarán, is another contender. Both have elevations above sea level more than 2 kilometres (6,600 ft) less than that of Everest.

INTEGRATED GEOGRAPHY

Rice terraces.

Environmental geography (also referred to as environmental geography, Integrated geography or human–environment geography) is the branch of geography that describes and explains the spatial aspects of interactions between human individuals or societies and their natural environment, these interactions being called coupled human–environment system. Summed up, environmental geography is about humans and nature and how we affect the environment and our planet.

It requires an understanding of the dynamics of physical geography, as well as the ways in which human societies conceptualize the environment (human geography). Thus, to a certain degree, it may be seen as a successor of Physische Anthropogeographie ("physical anthropogeography")—a term coined by University of Vienna geographer Albrecht Penck in 1924—and geographical cultural or human ecology. Integrated geography in the United States is principally influenced by the schools of Carl O. Sauer, whose perspective was rather historical, and Gilbert F. White , who developed a more applied view. Integrated geography (also, integrative geography, environmental geography or human–environment geography) is the branch of geography that describes and explains the spatial aspects of interactions between human individuals or societies and their natural environment, called coupled human–environment systems.

Focus

Wildlife refuge.

The links between human and physical geography were once more apparent than they are today. As human experience of the world is increasingly mediated by technology, the relationships between humans and the environment have often become obscured. Thereby, integrated geography represents a critically important set of analytical tools for assessing the impact of human presence on the environment. This is done by measuring the result of human activity on natural landforms and cycles. Methods for which this information is gained include remote sensing, and geographic information systems. Integrated geography helps us to ponder the environment in terms of its relationship to people. With integrated geography we can analyze different social science and humanities perspectives and their use in understanding people environment processes. Hence, it is considered the third branch of geography, the other branches being physical and human geography.

BIOGEOGRAPHY

Biogeography is the study of the distribution of species and ecosystems in geographic space and through geological time. Organisms and biological communities often vary in a regular fashion along geographic gradients of latitude, elevation, isolation and habitat area. Phytogeography is the branch of biogeography that studies the distribution of plants. Zoogeography is the branch that studies distribution of animals.

Knowledge of spatial variation in the numbers and types of organisms is as vital to us today as it was to our early human ancestors, as we adapt to heterogeneous but geographically predictable environments. Biogeography is an integrative field of inquiry that unites concepts and information from ecology, evolutionary biology, geology, and physical geography.

Modern biogeographic research combines information and ideas from many fields, from the physiological and ecological constraints on organismal dispersal to geological and climatological phenomena operating at global spatial scales and evolutionary time frames.

The short-term interactions within a habitat and species of organisms describe the ecological application of biogeography. Historical biogeography describes the long-term, evolutionary periods of time for broader classifications of organisms. Early scientists, beginning with Carl Linnaeus, contributed to the development of biogeography as a science. Beginning in the mid-18th century, Europeans explored the world and discovered the biodiversity of life.

The scientific theory of biogeography grows out of the work of Alexander von Humboldt, Hewett Cottrell Watson, Alphonse de Candolle, Alfred Russel Wallace, Philip Lutley Sclater and other biologists and explorers.

The patterns of species distribution across geographical areas can usually be explained through a combination of historical factors such as: speciation, extinction, continental drift, and glaciation. Through observing the geographic distribution of species, we can see associated variations in sea level, river routes, habitat, and river capture. Additionally, this science considers the geographic constraints of landmass areas and isolation, as well as the available ecosystem energy supplies.

Over periods of ecological changes, biogeography includes the study of plant and animal species in: their past and present living refugium habitat; their interim living sites; and their survival locales. Modern biogeography often employs the use of Geographic Information Systems (GIS), to understand the factors affecting organism distribution, and to predict future trends in organism distribution. Often mathematical models and GIS are employed to solve ecological problems that have a spatial aspect to them.

Biogeography is most keenly observed on the world's islands. These habitats are often much more manageable areas of study because they are more condensed than larger ecosystems on the mainland. Islands are also ideal locations because they allow scientists to look at habitats that new invasive species have only recently colonized and can observe how they disperse throughout the island and change it. They can then apply their understanding to similar but more complex mainland habitats. Islands are very diverse in their biomes, ranging from the tropical to arctic climates. This diversity in habitat allows for a wide range of species study in different parts of the world.

One scientist who recognized the importance of these geographic locations was Charles Darwin, who remarked in his journal "The Zoology of Archipelagoes will be well worth examination".

The scientific theory of biogeography grows out of the work of Alexander von Humboldt, Hewett Cottrell Watson, Alphonse de Candolle, Alfred Russel Wallace, Philip Lutley Sclater and other biologists and explorers.

The first discoveries that contributed to the development of biogeography as a science began in the mid-18th century, as Europeans explored the world and described the biodiversity of life. During the 18th century most views on the world were shaped around religion and for many natural theologists, the bible. Carl Linnaeus, in the mid-18th century, initiated the ways to classify organisms through his exploration of undiscovered territories. When he noticed that species were not as perpetual as he believed, he developed the Mountain Explanation to explain the distribution of biodiversity; when Noah's ark landed on Mount Ararat and the waters receded, the animals dispersed throughout different elevations on the mountain. This showed different species in different climates proving species were not constant. Linnaeus' findings set a basis for ecological biogeography. Through his strong beliefs in Christianity, he was inspired to classify the living world, which then gave way to additional accounts of secular views on geographical distribution. He argued that the structure of an animal was very closely related to its physical surroundings. This was important to a George Louis Buffon's rival theory of distribution.

Closely after Linnaeus, Georges-Louis Leclerc, Comte de Buffon observed shifts in climate and how species spread across the globe as a result. He was the first to see different groups of organisms in different regions of the world. Buffon saw similarities between some regions which led him to believe that at one point continents were connected and then water separated them and caused differences in species. His hypotheses were described by his books, Histoire Naturelle, and Générale et Particulière, in which he argued that varying geographical regions would have different forms of life. This was inspired by his observations comparing the Old and New World, as he determined distinct variations of species from the two regions. Buffon believed there was a single species creation event, and that different regions of the world were homes for varying species, which is an alternate view than that of Linnaeus. Buffon's law eventually became a principle of biogeography by explaining how similar environments were habitats for comparable types of organisms. Buffon also studied fossils which led him to believe that the earth was over tens of thousands of years old, and that humans had not lived there long in comparison to the age of the earth.

Following this period of exploration came the Age of Enlightenment in Europe, which attempted to explain the patterns of biodiversity observed by Buffon and Linnaeus. At the end of the 18th century, Alexander von Humboldt, known as the "founder of plant geography", developed the concept of physique generale to demonstrate the unity of science and how species fit together. As one of the first to contribute empirical data to the science of biogeography through his travel as an explorer, he observed differences in climate and vegetation. The earth was divided into regions which he defined as tropical, temperate, and arctic and within these regions there were similar forms of vegetation. This ultimately enabled him to create the isotherm, which allowed scientists to see patterns of life within different climates. He contributed his observations to findings of botanical geography by previous scientists, and sketched this description of both the biotic and abiotic features of the earth in his book, Cosmos.

Augustin de Candolle contributed to the field of biogeography as he observed species competition and the several differences that influenced the discovery of the diversity of life. He was a Swiss botanist and created the first Laws of Botanical Nomenclature in his work, Prodromus. He discussed plant distribution and his theories eventually had a great impact on Charles Darwin, who was inspired to consider species adaptations and evolution after learning about botanical geography. De Candolle was the first to describe the differences between the small-scale and large-scale distribution patterns of organisms around the globe.

In the 19th century, several additional scientists contributed new theories to further develop the concept of biogeography. Charles Lyell, being one of the first contributors in the 19th century, developed the Theory of Uniformitarianism after studying fossils. This theory explained how the world was not created by one sole catastrophic event, but instead from numerous creation events and locations. Uniformitarianism also introduced the idea that the Earth was actually significantly older than was previously accepted. Using this knowledge, Lyell concluded that it was possible for species to go extinct. Since he noted that earth's climate changes, he realized that species distribution must also change accordingly. Lyell argued that climate changes complemented vegetation changes, thus connecting the environmental surroundings to varying species. This largely influenced Charles Darwin in his development of the theory of evolution.

Charles Darwin was a natural theologist who studied around the world, and most importantly in the Galapagos Islands. Darwin introduced the idea of natural selection, as he theorized against previously accepted ideas that species were static or unchanging. His contributions to biogeography and the theory of evolution were different from those of other explorers of his time, because he developed a mechanism to describe the ways that species changed. His influential ideas include the development of theories regarding the struggle for existence and natural selection. Darwin's theories started a biological segment to biogeography and empirical studies, which enabled future scientists to develop ideas about the geographical distribution of organisms around the globe.

Alfred Russel Wallace studied the distribution of flora and fauna in the Amazon Basin and the Malay Archipelago in the mid-19th century. His research was essential to the further development of biogeography, and he was later nicknamed the "father of Biogeography". Wallace conducted fieldwork researching the habits, breeding and migration tendencies, and feeding behavior of thousands of species. He studied butterfly and bird distributions in comparison to the presence or absence of geographical barriers. His observations led him to conclude that the number of organisms present in a community was dependent on the amount of food resources in the particular habitat. Wallace believed species were dynamic by responding to biotic and abiotic factors. He and Philip Sclater saw biogeography as a source of support for the theory of evolution as they used Darwin's conclusion to explain how biogeography was similar to a record of species inheritance. Key findings, such as the sharp difference in fauna either side of the Wallace Line, and the sharp difference that existed between North and South America prior to their relatively recent faunal interchange, can only be understood in this light. Otherwise, the field of biogeography would be seen as a purely descriptive one.

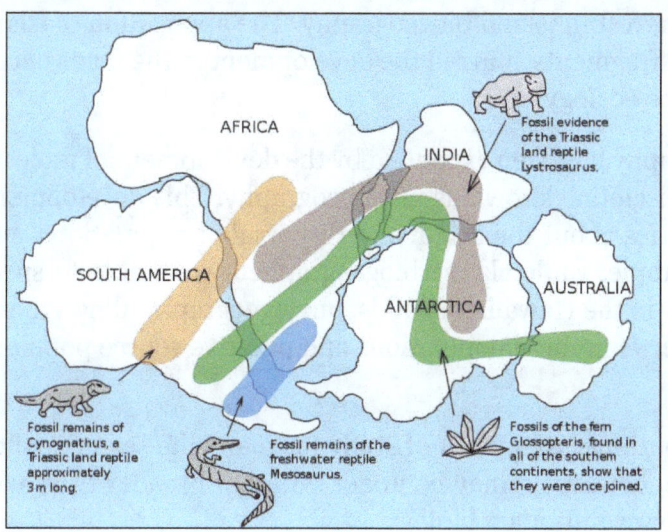

Moving on to the 20th century, Alfred Wegener introduced the Theory of Continental Drift in 1912, though it was not widely accepted until the 1960s. This theory was revolutionary because it changed the way that everyone thought about species and their distribution around the globe. The theory explained how continents were formerly joined together in one large landmass, Pangea, and slowly drifted apart due to the movement of the plates below Earth's surface. The evidence for this theory is in the geological similarities between varying locations around the globe, fossil comparisons from different continents, and the jigsaw puzzle shape of the landmasses on Earth. Though Wegener did not know the mechanism of this concept of Continental Drift, this contribution to the study of biogeography was significant in the way that it shed light on the importance of environmental and geographic similarities or differences as a result of climate and other pressures on the planet. Importantly, late in his career Wegener recognised that testing his theory required measurement of continental movement rather than inference from fossils species distributions.

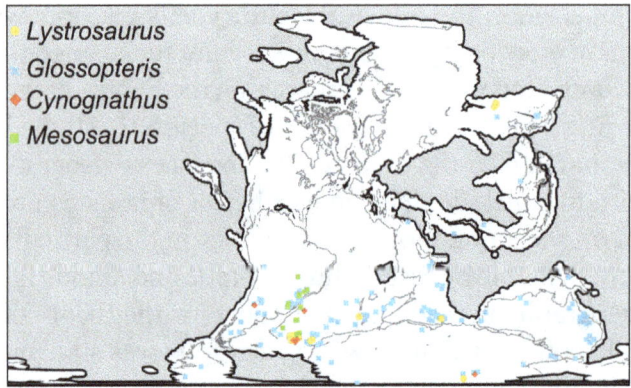

The publication of The Theory of Island Biogeography by Robert MacArthur and E.O. Wilson in 1967 showed that the species richness of an area could be predicted in terms of such factors as habitat area, immigration rate and extinction rate. This added to the long-standing interest in island biogeography. The application of island biogeography theory to habitat fragments spurred the development of the fields of conservation biology and landscape ecology.

Classic biogeography has been expanded by the development of molecular systematics, creating a new discipline known as phylogeography. This development allowed scientists to test theories about the origin and dispersal of populations, such as island endemics. For example, while classic biogeographers were able to speculate about the origins of species in the Hawaiian Islands, phylogeography allows them to test theories of relatedness between these populations and putative source populations in Asia and North America.

Biogeography continues as a point of study for many life sciences and geography students worldwide, however it may be under different broader titles within institutions such as ecology or evolutionary biology.

In recent years, one of the most important and consequential developments in biogeography has been to show how multiple organisms, including mammals like monkeys and reptiles like lizards, overcame barriers such as large oceans that many biogeographers formerly believed were impossible to cross.

Modern Applications

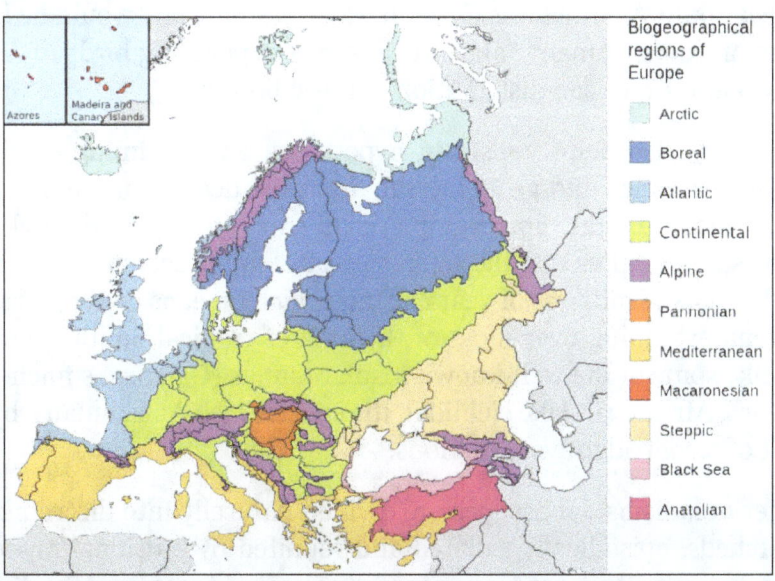

Biogeography now incorporates many different fields including but not limited to physical geography, geology, botany and plant biology, zoology, general biology, and modelling. A biogeographer's main focus is on how the environment and humans affect the distribution of species as well as other manifestations of Life such as species or genetic diversity. Biogeography is being applied to biodiversity conservation and planning, to projecting global environmental changes on species and biomes, to projecting the spread of infectious diseases, invasive species, and for supporting planning for the establishment of crops. Technological advances have allowed generating a whole suit of predictor variables for biogeographic analysis, including satellite imaging and processing of the Earth. Two main types of satellite imaging that are important within modern biogeography are Global Production Efficiency Model (GLO-PEM) and Geographic Information Systems (GIS). GLO-PEM uses satellite-imaging gives "repetitive, spatially contiguous, and time specific observations of vegetation". These observations are on a global scale. GIS can show certain processes on the earth's surface like whale locations, sea surface temperatures, and bathymetry. Current scientists also use coral reefs to delve into the history of biogeography through the fossilized reefs.

Paleobiogeography

Paleobiogeography goes one step further to include palaeogeographic data and considerations of plate tectonics. Using molecular analyses and corroborated by fossils, it has been possible to demonstrate that perching birds evolved first in the region of Australia or the adjacent Antarctic (which at that time lay somewhat further north and had a temperate climate). From there, they spread to the other Gondwanan continents and Southeast Asia – the part of Laurasia then closest to their origin of dispersal – in the late Paleogene, before achieving a global distribution in the early Neogene. Not knowing that at the time of dispersal, the Indian Ocean was much narrower than it is today, and that South America was closer to the Antarctic, one would be hard pressed to explain the presence of many "ancient" lineages of perching birds in Africa, as well as the mainly South American distribution of the suboscines.

Paleobiogeography also helps constrain hypotheses on the timing of biogeographic events such as vicariance and geodispersal, and provides unique information on the formation of regional biotas. For example, data from species-level phylogenetic and biogeographic studies tell us that the Amazonian fish fauna accumulated in increments over a period of tens of millions of years, principally by means of allopatric speciation, and in an arena extending over most of the area of tropical South America. In other words, unlike some of the well-known insular faunas (Galapagos finches, Hawaiian drosophilid flies, African rift lake cichlids), the species-rich Amazonian ichthyofauna is not the result of recent adaptive radiations.

For freshwater organisms, landscapes are divided naturally into discrete drainage basins by watersheds, episodically isolated and reunited by erosional processes. In regions like the Amazon Basin (or more generally Greater Amazonia, the Amazon basin, Orinoco basin, and Guianas) with an exceptionally low (flat) topographic relief, the many waterways have had a highly reticulated history over geological time. In such a context, stream capture is an important factor affecting the evolution and distribution of freshwater organisms. Stream capture occurs when an upstream portion of one river drainage is diverted to the downstream portion of an adjacent basin. This can happen as a result of tectonic uplift (or subsidence), natural damming created by a landslide, or headward or lateral erosion of the watershed between adjacent basins.

Concepts and Fields

Biogeography is a synthetic science, related to geography, biology, soil science, geology, climatology, ecology and evolution.

Some fundamental concepts in biogeography include:

- Allopatric speciation: The splitting of a species by evolution of geographically isolated populations.

- Evolution: Change in genetic composition of a population.

- Extinction: Disappearance of a species.

- Dispersal: Movement of populations away from their point of origin, related to migration.

- Endemic areas.

- Geodispersal: The erosion of barriers to biotic dispersal and gene flow, that permit range expansion and the merging of previously isolated biotas.

- Range and distribution.

- Vicariance: The formation of barriers to biotic dispersal and gene flow, that tend to subdivide species and biotas, leading to speciation and extinction; vicariance biogeography is the field that studies these patterns.

Comparative Biogeography

The study of comparative biogeography can follow two main lines of investigation:

- Systematic biogeography: The study of biotic area relationships, their distribution, and hierarchical classification.

- Evolutionary biogeography: The proposal of evolutionary mechanisms responsible for organismal distributions. Possible mechanisms include widespread taxa disrupted by continental break-up or individual episodes of long-distance movement.

Biogeographic Regionalisations

There are many types of biogeographic units used in biogeographic regionalisation schemes, as there are many criteria (species composition, physiognomy, ecological aspects) and hierarchization schemes: biogeographic realms (or ecozones), bioregions (sensu stricto), ecoregions, zoogeographical regions, floristic regions, vegetation types, biomes, etc.

The terms biogeographic unit, biogeographic area or bioregion sensu lato, can be used for these categories, regardless of rank.

Recently, an International Code of Area Nomenclature was proposed for biogeography.

GEOMORPHOLOGY

Geomorphology is the scientific study of the origin and evolution of topographic and bathymetric features created by physical, chemical or biological processes operating at or near the Earth's surface. Geomorphologists seek to understand why landscapes look the way they do, to understand landform history and dynamics and to predict changes through a combination of field observations, physical experiments and numerical modeling. Geomorphologists work within disciplines such as physical geography, geology, geodesy, engineering geology, archaeology, climatology and geotechnical engineering. This broad base of interests contributes to many research styles and interests within the field.

Earth's surface is modified by a combination of surface processes that shape landscapes, and geologic processes that cause tectonic uplift and subsidence, and shape the coastal geography. Surface processes comprise the action of water, wind, ice, fire, and living things on the surface of the Earth, along with chemical reactions that form soils and alter material properties, the stability and rate of change of topography under the force of gravity, and other factors, such as (in the very recent past) human alteration of the landscape. Many of these factors are strongly mediated by climate. Geologic processes include the uplift of mountain ranges, the growth of volcanoes, isostatic changes in land surface elevation (sometimes in response to surface processes), and the formation of deep sedimentary basins where the surface of the Earth drops and is filled with material eroded from other parts of the landscape. The Earth's surface and its topography therefore are an intersection of climatic, hydrologic, and biologic action with geologic processes, or alternatively stated, the intersection of the Earth's lithosphere with its hydrosphere, atmosphere, and biosphere.

The broad-scale topographies of the Earth illustrate this intersection of surface and subsurface action. Mountain belts are uplifted due to geologic processes. Denudation of these high uplifted regions produces sediment that is transported and deposited elsewhere within the landscape or off the coast. On progressively smaller scales, similar ideas apply, where individual landforms evolve in response to the balance of additive processes (uplift and deposition) and subtractive processes (subsidence and erosion). Often, these processes directly affect each other: ice sheets, water, and sediment are all loads that change topography through flexural isostasy. Topography can modify the local climate, for example through orographic precipitation, which in turn modifies the topography by changing the hydrologic regime in which it evolves. Many geomorphologists are particularly interested in the potential for feedbacks between climate and tectonics, mediated by geomorphic processes.

In addition to these broad-scale questions, geomorphologists address issues that are more specific and more local. Glacial geomorphologists investigate glacial deposits such as moraines, eskers, and proglacial lakes, as well as glacial erosional features, to build chronologies of both small glaciers and large ice sheets and understand their motions and effects upon the landscape. Fluvial geomorphologists focus on rivers, how they transport sediment, migrate across the landscape, cut into bedrock, respond to environmental and tectonic changes, and interact with humans. Soils geomorphologists investigate soil profiles and chemistry to learn about the history of a particular landscape and understand how climate, biota, and rock interact. Other geomorphologists study how hillslopes form and change. Still others investigate the relationships between ecology and geomorphology. Because geomorphology is defined to comprise everything related to the surface of the Earth and its modification, it is a broad field with many facets.

Geomorphologists use a wide range of techniques in their work. These may include fieldwork and field data collection, the interpretation of remotely sensed data, geochemical analyses, and the numerical modelling of the physics of landscapes. Geomorphologists may rely on geochronology, using dating methods to measure the rate of changes to the surface. Terrain measurement techniques are vital to quantitatively describe the form of the Earth's surface, and include differential GPS, remotely sensed digital terrain models and laser scanning, to quantify, study, and to generate illustrations and maps.

Practical applications of geomorphology include hazard assessment (such as landslide prediction and mitigation), river control and stream restoration, and coastal protection. Planetary geomorphology studies landforms on other terrestrial planets such as Mars. Indications of effects of wind, fluvial, glacial, mass wasting, meteor impact, tectonics and volcanic processes are studied. This effort not only helps better understand the geologic and atmospheric history of those planets but also extends geomorphological study of the Earth. Planetary geomorphologists often use Earth analogues to aid in their study of surfaces of other planets.

Other than some notable exceptions in antiquity, geomorphology is a relatively young science, growing along with interest in other aspects of the earth sciences in the mid-19th century. This topic provides a very brief outline of some of the major figures and events in its development.

Ancient Geomorphology

The study of landforms and the evolution of the Earth's surface can be dated back to scholars of Classical Greece. Herodotus argued from observations of soils that the Nile delta was actively growing into the Mediterranean Sea, and estimated its age. Aristotle speculated that due to sediment transport into the sea, eventually those seas would fill while the land lowered. He claimed that this would mean that land and water would eventually swap places, whereupon the process would begin again in an endless cycle.

Another early theory of geomorphology was devised by the polymath Chinese scientist and statesman Shen Kuo. This was based on his observation of marine fossil shells in a geological stratum of a mountain hundreds of miles from the Pacific Ocean.

Noticing bivalve shells running in a horizontal span along the cut section of a cliff-side, he theorized that the cliff was once the pre-historic location of a seashore that had shifted hundreds of miles over the centuries. He inferred that the land was re-shaped and formed by soil erosion of the mountains and by deposition of silt, after observing strange natural erosions of the Taihang Mountains and the Yandang Mountain near Wenzhou. Furthermore, he promoted the theory of gradual climate change over centuries of time once ancient petrified bamboos were found to be preserved underground in the dry, northern climate zone of Yanzhou, which is now modern day Yan'an, Shaanxi province.

Early Modern Geomorphology

The term geomorphology seems to have been first used by Laumann in an 1858 work written in German. Keith Tinkler has suggested that the word came into general use in English, German and French after John Wesley Powell and W. J. McGee used it during the International Geological Conference of 1891. John Edward Marr in his The Scientific Study of Scenery considered his book as, 'an Introductory Treatise on Geomorphology, a subject which has sprung from the union of Geology and Geography'.

An early popular geomorphic model was the geographical cycle or cycle of erosion model of broad-scale landscape evolution developed by William Morris Davis between 1884 and 1899. It was an elaboration of the uniformitarianism theory that had first been proposed by James Hutton With regard to valley forms, for example, uniformitarianism posited a sequence in which a river runs through a flat terrain, gradually carving an increasingly deep valley, until the side valleys eventually erode, flattening the terrain again, though at a lower elevation. It was thought that tectonic uplift could then start the cycle over. In the decades following Davis's development of this idea, many of those studying geomorphology sought to fit their findings into this framework, known today as "Davisian". Davis's ideas are of historical importance, but have been largely superseded today, mainly due to their lack of predictive power and qualitative nature.

In the 1920s, Walther Penck developed an alternative model to Davis's. Penck thought that landform evolution was better described as an alternation between ongoing processes of uplift and denudation, as opposed to Davis's model of a single uplift followed by decay. He also emphasised that in many landscapes slope evolution occurs by back-wearing of rocks, not by Davisian-style surface lowering, and his science tended to emphasise surface process over understanding in detail the surface history of a given locality. Penck was German, and during his lifetime his ideas were at times rejected vigorously by the English-speaking geomorphology community. His early death, Davis' dislike for his work, and his at-times-confusing writing style likely all contributed to this rejection.

Both Davis and Penck were trying to place the study of the evolution of the Earth's surface on a more generalized, globally relevant footing than it had been previously. In the early 19th century, authors – especially in Europe – had tended to attribute the form of landscapes to local climate, and in particular to the specific effects of glaciation and periglacial processes. In contrast, both Davis and Penck were seeking to emphasize the importance of evolution of landscapes through time and the generality of the Earth's surface processes across different landscapes under different conditions.

During the early 1900s, the study of regional-scale geomorphology was termed "physiography". Physiography later was considered to be a contraction of "physical" and "geography", and therefore synonymous with physical geography, and the concept became embroiled in controversy surrounding the appropriate concerns of that discipline. Some geomorphologists held to a geological basis for physiography and emphasized a concept of physiographic regions while a conflicting trend among geographers was to equate physiography with "pure morphology", separated from its geological heritage. In the period following World War II, the emergence of process, climatic, and quantitative studies led to a preference by many earth scientists for the term "geomorphology" in order to suggest an analytical approach to landscapes rather than a descriptive one.

Climatic Geomorphology

During the age of New Imperialism in the late 19th century European explorers and scientists traveled across the globe bringing descriptions of landscapes and landforms. As geographical knowledge increased over time these observations were systematized in a search for regional patterns. Climate emerged thus as prime factor for explaining landform distribution at a grand scale. The rise of climatic geomorphology was foreshadowed by the work of Wladimir Köppen, Vasily Dokuchaev and Andreas Schimper. William Morris Davis, the leading geomorphologist of his time, recognized the role of climate by complementing his "normal" temperate climate cycle of erosion with arid and glacial ones. Nevertheless, interest in climatic geomorphology was also a reaction against Davisian geomorphology that was by the mid-20th century considered both un-innovative and dubious. Early climatic geomorphology developed primarily in continental Europe while in the English-speaking world the tendency was not explicit until L.C. Peltier's 1950 publication on a periglacial cycle of erosion.

Climatic geomorphology was criticized in a 1969 review article by process geomorphologist D.R. Stoddart. The criticism by Stoddart proved "devastating" sparking a decline in the popularity of climatic geomorphology in the late 20th century. Stoddart criticized climatic geomorphology for applying supposedly "trivial" methodologies in establishing landform differences between morphoclimatic zones, being linked to Davisian geomorphology and by allegedly neglecting the fact that physical laws governing processes are the same across the globe. In addition some conceptions of climatic geomorphology, like that which holds that chemical weathering is more rapid in tropical climates than in cold climates proved to not be straightforwardly true.

Quantitative and Process Geomorphology

Part of the Great Escarpment in the Drakensberg, southern Africa. This
landscape, with its high altitude plateau being incised into by the steep slopes
of the escarpment, was cited by Davis as a classic example of his cycle of erosion.

Geomorphology was started to be put on a solid quantitative footing in the middle of
the 20th century. Following the early work of Grove Karl Gilbert around the turn of the
20th century, a group of mainly American natural scientists, geologists and hydraulic
engineers including William Walden Rubey, Ralph Alger Bagnold, Hans Albert Ein-
stein, Frank Ahnert, John Hack, Luna Leopold, A. Shields, Thomas Maddock, Arthur
Strahler, Stanley Schumm, and Ronald Shreve began to research the form of landscape
elements such as rivers and hillslopes by taking systematic, direct, quantitative measurements of aspects of them and investigating the scaling of these measurements.
These methods began to allow prediction of the past and future behavior of landscapes
from present observations, and were later to develop into the modern trend of a highly quantitative approach to geomorphic problems. Many groundbreaking and widely
cited early geomorphology studies appeared in the Bulletin of the Geological Society of
America, and received only few citations prior to 2000 (they are examples of "sleeping
beauties") when a marked increase in quantitative geomorphology research occurred.

Quantitative geomorphology can involve fluid dynamics and solid mechanics, geomor-
phometry, laboratory studies, field measurements, theoretical work, and full landscape
evolution modeling. These approaches are used to understand weathering and the formation of soils, sediment transport, landscape change, and the interactions between
climate, tectonics, erosion, and deposition.

In Sweden Filip Hjulström's doctoral thesis, "The River Fyris", contained one of the first
quantitative studies of geomorphological processes ever published. His students followed in the same vein, making quantitative studies of mass transport (Anders Rapp),
fluvial transport (Åke Sundborg), delta deposition (Valter Axelsson), and coastal processes (John O. Norrman). This developed into "the Uppsala School of Physical Geography".

Contemporary Geomorphology

Today, the field of geomorphology encompasses a very wide range of different approaches and interests. Modern researchers aim to draw out quantitative "laws" that govern Earth surface processes, but equally, recognize the uniqueness of each landscape and environment in which these processes operate. Particularly important realizations in contemporary geomorphology include:

- That not all landscapes can be considered as either "stable" or "perturbed", where this perturbed state is a temporary displacement away from some ideal target form. Instead, dynamic changes of the landscape are now seen as an essential part of their nature.

- That many geomorphic systems are best understood in terms of the stochasticity of the processes occurring in them, that is, the probability distributions of event magnitudes and return times. This in turn has indicated the importance of chaotic determinism to landscapes, and that landscape properties are best considered statistically. The same processes in the same landscapes do not always lead to the same end results.

Albeit having its importance diminished climatic geomorphology continues to exist as field of study producing relevant research. More recently concerns over global warming have led to a renewed interest in the field.

Despite considerable criticism the cycle of erosion model has remained part of the science of geomorphology. The model or theory has never been proved wrong, but neither has it been proven. The inherent difficulties of the model have instead made geomorphological research to advance along other lines. In contrast to its disputed status in geomorphology, the cycle of erosion model is a common approach used to establish denudation chronologies, and is thus an important concept in the science of historical geology. While acknowledging its shortcomings modern geomorphologists Andrew Goudie and Karna Lidmar-Bergström have praised it for its elegance and pedagogical value respectively.

Processes

Geomorphically relevant processes generally fall into (1) the production of regolith by weathering and erosion, (2) the transport of that material, and (3) its eventual deposition. Primary surface processes responsible for most topographic features include wind, waves, chemical dissolution, mass wasting, groundwater movement, surface water flow, glacial action, tectonism, and volcanism. Other more exotic geomorphic processes might include periglacial (freeze-thaw) processes, salt-mediated action, changes to the seabed caused by marine currents, seepage of fluids through the seafloor or extraterrestrial impact.

Gorge cut by the Indus river into bedrock, Nanga Parbat region, Pakistan.
This is the deepest river canyon in the world. Nanga Parbat itself,
the world's 9th highest mountain, is seen in the background.

Aeolian Processes

Wind-eroded alcove.

Aeolian processes pertain to the activity of the winds and more specifically, to the winds' ability to shape the surface of the Earth. Winds may erode, transport, and deposit materials, and are effective agents in regions with sparse vegetation and a large supply of fine, unconsolidated sediments. Although water and mass flow tend to mobilize more material than wind in most environments, aeolian processes are important in arid environments such as deserts.

Biological Processes

The interaction of living organisms with landforms, or biogeomorphologic processes, can be of many different forms, and is probably of profound importance for the terrestrial geomorphic system as a whole. Biology can influence very many geomorphic processes, ranging from biogeochemical processes controlling chemical weathering, to the influence of mechanical processes like burrowing and tree throw on soil development, to even controlling global erosion rates through modulation of climate through carbon dioxide balance. Terrestrial landscapes in which the role of biology in mediating surface processes can be definitively excluded are extremely rare, but may hold important information for understanding the geomorphology of other planets, such as Mars.

Beaver dams, as this one in Tierra del Fuego, constitute a specific form of zoogeomorphology, a type of biogeomorphology.

Fluvial Processes

Seif and barchan dunes in the Hellespontus region on the surface of Mars. Dunes are mobile landforms created by the transport of large volumes of sand by wind.

Rivers and streams are not only conduits of water, but also of sediment. The water, as it flows over the channel bed, is able to mobilize sediment and transport it downstream, either as bed load, suspended load or dissolved load. The rate of sediment transport depends on the availability of sediment itself and on the river's discharge. Rivers are also capable of eroding into rock and creating new sediment, both from their own beds and also by coupling to the surrounding hillslopes. In this way, rivers are thought of as setting the base level for large-scale landscape evolution in nonglacial environments. Rivers are key links in the connectivity of different landscape elements.

As rivers flow across the landscape, they generally increase in size, merging with other rivers. The network of rivers thus formed is a drainage system. These systems take on four general patterns: dendritic, radial, rectangular, and trellis. Dendritic happens to be the most common, occurring when the underlying stratum is stable (without fault-ing). Drainage systems have four primary components: drainage basin, alluvial valley, delta plain, and receiving basin. Some geomorphic examples of fluvial landforms are alluvial fans, oxbow lakes, and fluvial terraces.

Glacial Processes

Glaciers, while geographically restricted, are effective agents of landscape change. The gradual movement of ice down a valley causes abrasion and plucking of the under-lying rock. Abrasion produces fine sediment, termed glacial flour. The debris trans-ported by the glacier, when the glacier recedes, is termed a moraine. Glacial erosion is responsible for U-shaped valleys, as opposed to the V-shaped valleys of fluvial origin.

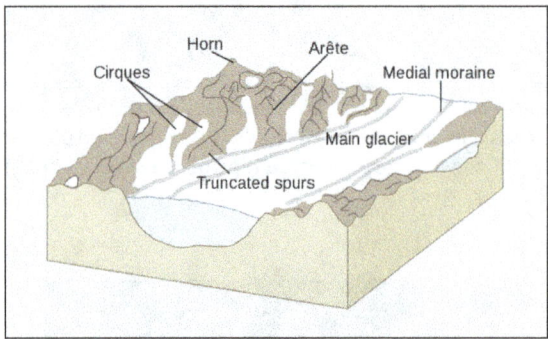

Features of a glacial landscape.

The way glacial processes interact with other landscape elements, particularly hillslope and fluvial processes, is an important aspect of Plio-Pleistocene landscape evolution and its sedimentary record in many high mountain environments. Environments that have been relatively recently glaciated but are no longer may still show elevated land-scape change rates compared to those that have never been glaciated. Nonglacial geo-morphic processes which nevertheless have been conditioned by past glaciation are termed paraglacial processes. This concept contrasts with periglacial processes, which are directly driven by formation or melting of ice or frost.

Hillslope Processes

Talus cones on the north shore of Isfjorden, Svalbard, Norway. Talus cones are accumulations of coarse hillslope debris at the foot of the slopes producing the material.

The Ferguson Slide is an active landslide in the Merced River canyon
on California State Highway 140, a primary access road to Yosemite National Park.

Soil, regolith, and rock move downslope under the force of gravity via creep, slides, flows, topples, and falls. Such mass wasting occurs on both terrestrial and submarine slopes, and has been observed on Earth, Mars, Venus, Titan and Iapetus.

Ongoing hillslope processes can change the topology of the hillslope surface, which in turn can change the rates of those processes. Hillslopes that steepen up to certain critical thresholds are capable of shedding extremely large volumes of material very quickly, making hillslope processes an extremely important element of landscapes in tectonically active areas.

On the Earth, biological processes such as burrowing or tree throw may play important roles in setting the rates of some hillslope processes.

Igneous Processes

Both volcanic (eruptive) and plutonic (intrusive) igneous processes can have important impacts on geomorphology. The action of volcanoes tends to rejuvenize landscapes, covering the old land surface with lava and tephra, releasing pyroclastic material and forcing rivers through new paths. The cones built by eruptions also build substantial new topography, which can be acted upon by other surface processes. Plutonic rocks intruding then solidifying at depth can cause both uplift or subsidence of the surface, depending on whether the new material is denser or less dense than the rock it displaces.

Tectonic Processes

Tectonic effects on geomorphology can range from scales of millions of years to minutes or less. The effects of tectonics on landscape are heavily dependent on the nature of the underlying bedrock fabric that more or less controls what kind of local morphology tectonics can shape. Earthquakes can, in terms of minutes, submerge large areas of land creating new wetlands. Isostatic rebound can account for significant changes over hundreds to thousands of years, and allows erosion of a mountain belt to promote further erosion as mass is removed from the chain and the belt uplifts. Long-term plate tectonic dynamics give rise to orogenic belts, large mountain chains with typical lifetimes of many tens of millions of years, which form focal points for high rates of fluvial and hillslope processes and thus long-term sediment production.

Features of deeper mantle dynamics such as plumes and delamination of the lower lithosphere have also been hypothesised to play important roles in the long term (> million year), large scale (thousands of km) evolution of the Earth's topography. Both can promote surface uplift through isostasy as hotter, less dense, mantle rocks displace cooler, denser, mantle rocks at depth in the Earth.

Marine Processes

Marine processes are those associated with the action of waves, marine currents and seepage of fluids through the seafloor. Mass wasting and submarine landsliding are also important processes for some aspects of marine geomorphology. Because ocean basins are the ultimate sinks for a large fraction of terrestrial sediments, depositional processes and their related forms (e.g., sediment fans, deltas) are particularly important as elements of marine geomorphology.

Scales

Different geomorphological processes dominate at different spatial and temporal scales. Moreover, scales on which processes occur may determine the reactivity or otherwise of landscapes to changes in driving forces such as climate or tectonics. These ideas are key to the study of geomorphology today.

To help categorize landscape scales some geomorphologists might use the following taxonomy:

- 1st: Continent, ocean basin, climatic zone ($\sim$10,000,000 km^2).

- 2nd: Shield, e.g. Baltic Shield, or mountain range ($\sim$1,000,000 km^2).

- 3rd: Isolated sea, Sahel ($\sim$100,000 km^2).

- 4th: Massif, e.g. Massif Central or Group of related landforms, e.g., Weald ($\sim$10,000 km^2).

- 5th: River valley, Cotswolds ($\sim$1,000 km^2).

- 6th: Individual mountain or volcano, small valleys ($\sim$100 km^2).

- 7th: Hillslopes, stream channels, estuary ($\sim$10 km^2).

- 8th: Gully, barchannel ($\sim$1 km^2).

- 9th: Meter-sized features.

Overlap with other Fields

There is a considerable overlap between geomorphology and other fields. Deposition of material is extremely important in sedimentology. Weathering is the chemical and physical disruption of earth materials in place on exposure to atmospheric or near surface agents, and is typically studied by soil scientists and environmental chemists, but is an essential component of geomorphology because it is what provides the material that can be moved in the first place. Civil and environmental engineers are concerned with erosion and sediment transport, especially related to canals, slope stability (and natural hazards), water quality, coastal environmental management, transport of contaminants, and stream restoration. Glaciers can cause extensive erosion and deposition in a short period of time, making them extremely important entities in the high latitudes and meaning that they set the conditions in the headwaters of mountain-born streams; glaciology therefore is important in geomorphology.

CLIMATOLOGY

Climatology or climate science is the scientific study of climate, scientifically defined as weather conditions averaged over a period of time. This modern field of study is regarded as a branch of the atmospheric sciences and a subfield of physical geography, which is one of the Earth sciences. Climatology now includes aspects of oceanography and biogeochemistry.

The main methods employed by climatologists are the analysis of observations and modelling the physical laws that determine the climate. The main topics of research are the study of climate variability, mechanisms of climate changes and modern climate change. Basic knowledge of climate can be used within shorter term weather forecasting, for instance about climatic cycles such as the El Niño–Southern Oscillation (ENSO), the Madden–Julian oscillation (MJO), the North Atlantic oscillation (NAO), the Arctic oscillation (AO), the Pacific decadal oscillation (PDO), and the Interdecadal Pacific Oscillation (IPO).

Climate models are used for a variety of purposes from study of the dynamics of the weather and climate system to projections of future climate. Weather is known as the condition of the atmosphere over a period of time, while climate has to do with the atmospheric condition over an extended to indefinite period of time.

Arguably the most influential classic text on climate was On Airs, Water and Places written by Hippocrates around 400 BCE. This work commented on the effect of climate on human health and cultural differences between Asia and Europe. This idea that climate controls which countries excel depending on their climate, or climatic determinism, remained influential throughout history. Chinese scientist Shen Kuo inferred that climates naturally shifted over an enormous span of time, after observing petrified bamboos found underground near Yanzhou (modern day Yan'an, Shaanxi province), a dry-climate area unsuitable for the growth of bamboo.

The invention of the thermometer and the barometer during the Scientific Revolution allowed for systematic recordkeeping, that began as early as 1640 in England. Early climate researchers include Edmund Halley, who published a map of the trade winds in 1686 after a voyage to the southern hemisphere. Benjamin Franklin first mapped the course of the Gulf Stream for use in sending mail from the United States to Europe. Francis Galton invented the term anticyclone. Helmut Landsberg fostered the use of statistical analysis in climatology, which led to its evolution into a physical science.

In the early 20th century, climatology was mostly focused on the description of re gional climates. This descriptive climatology was mainly an applied science, giving farmers and other interested people statistics about what the normal weather was and how big chances were of extreme events. To do this, climatologists had to define a climate normal, or an average of weather and weather extremes over a period of typically 30 years.

Around the middle of the 20th century, many assumptions in meteorology and climatology considered climate to be roughly constant. While scientists knew of past climate change such as the ice ages, the concept of climate as unchanging was useful in the development of a general theory of what determines climate. This started to change in the decades that followed, and while the history of climate change science started earlier, climate change only became one of the mean topics of study for climatologists in the seventies and onward.

Subfields

Various subfields of climatology study different aspects of the climate. There are different categorizations of the fields in climatology. The American Meteorological Society for instance identifies descriptive climatology, scientific climatology and applied climatology as the three subcategories of climatology, a categorization based on the complexity and the purpose of the research. Applied climatologists apply their expertise to different industries such as manufacturing and agriculture.

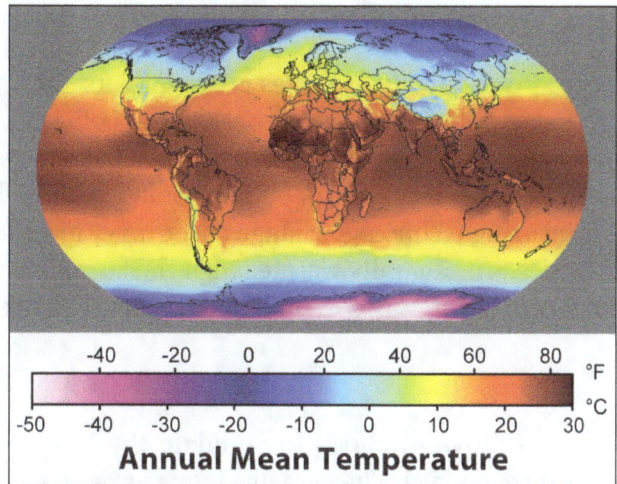

-40   -20   0   20   40   60   80   °F

°C
-50   -40   -30   -20   -10   0   10   20   30

Annual Mean Temperature

Map of the average temperature over 30 years. Data sets formed from the long-term average of historical weather parameters are sometimes called a "climatology".

Paleoclimatology seeks to reconstruct and understand past climates by examining records such as ice cores and tree rings (dendroclimatology). Paleotempestology uses these same records to help determine hurricane frequency over millennia. Historical climatology is the study of climate as related to human history and thus focuses only on the last few thousand years.

Boundary-layer climatology is preoccupied with exchanges in water, energy and momentum near the surface. Further identified subfields are physical climatology, dynamic climatology, tornado climatology, regional climatology, bioclimatology, applied climatology, and synoptic climatology. The study of the hydrological cycle at long time scales (hydroclimatology) is further subdivided within the subfields of snow climatology and hail climatology.

Methods

The study of contemporary climates incorporates meteorological data accumulated over many years, such as records of rainfall, temperature and atmospheric composition. Knowledge of the atmosphere and its dynamics is also embodied in models, either statistical or mathematical, which help by integrating different observations and testing how they fit together. Modeling is used for understanding past, present and potential future climates.

Climate research is made difficult by the large scale, long time periods, and complex processes which govern climate. Climate is governed by physical laws which can be expressed as differential equations. These equations are coupled and nonlinear, so that approximate solutions are obtained by using numerical methods to create global climate models. Climate is sometimes modeled as a stochastic process but this is generally accepted as an approximation to processes that are otherwise too complicated to analyze.

Climate Data

The collection of long record of climate variables is essential for the study of climate. Climatology deals with the aggregate data that meteorology has collected. Scientists use both direct and indirect observations of the climate, from Earth observing satellites and scientific instrumentation such as a global network of thermometers, to prehistoric ice extracted from glaciers. As measuring technology changes over time, records of data cannot be compared directly. As cities are generally warmer than the surrounding areas, urbanization has made it necessary to constantly correct data for this urban heat island effect.

Models

Climate models use quantitative methods to simulate the interactions of the atmosphere, oceans, land surface, and ice. They are used for a variety of purposes from study of the dynamics of the weather and climate system to projections of future climate. All climate models balance, or very nearly balance, incoming energy as short wave (including visible) electromagnetic radiation to the earth with outgoing energy as long wave (infrared) electromagnetic radiation from the earth. Any unbalance results in a change in the average temperature of the earth. Most climate models include the radiative effects of greenhouse gases such as carbon dioxide. These models predict an upward trend in the surface temperatures, as well as a more rapid increase in temperature at higher latitudes.

Models can range from relatively simple to complex:

- A simple radiant heat transfer model that treats the earth as a single point and averages outgoing energy.

- this can be expanded vertically (radiative-convective models), or horizontally.

- Coupled atmosphere–ocean–sea ice global climate models discretise and solve the full equations for mass and energy transfer and radiant exchange.

- Earth system models further include the biosphere.

Topics that climatologists study fall roughly into three categories: climate variability, mechanisms of climate change and modern climate change.

Climatological Processes

Various factors impact the average state of the atmosphere at a particular location. For instance, midlatitudes will have a pronounced seasonal cycle in temperature whereas tropical regions show little variation in temperature over the year. Another major control in climate is continentality: the distance to major water bodies such as oceans. Oceans act as a moderating factor, so that land close to it has typically has mild winters and moderate summers. The atmosphere interacts with other spheres of the climate system, with winds generating ocean currents that transport heat around the globe.

Climate Classification

Classification is an important aspect of many sciences as a tool of simplifying complicated processes. Different climate classifications have been developed over the centuries, with the first ones in Ancient Greece. How climates are classified depends on what the application is. A wind energy producer will require different information (wind) in the classification than somebody interested in agriculture, for who precipitation and temperature are more important. The most widely used classification, the Köppen climate classification, was developed in the late nineteenth century and is based on vegetation. It uses monthly temperature and precipitation data.

Climate Variability

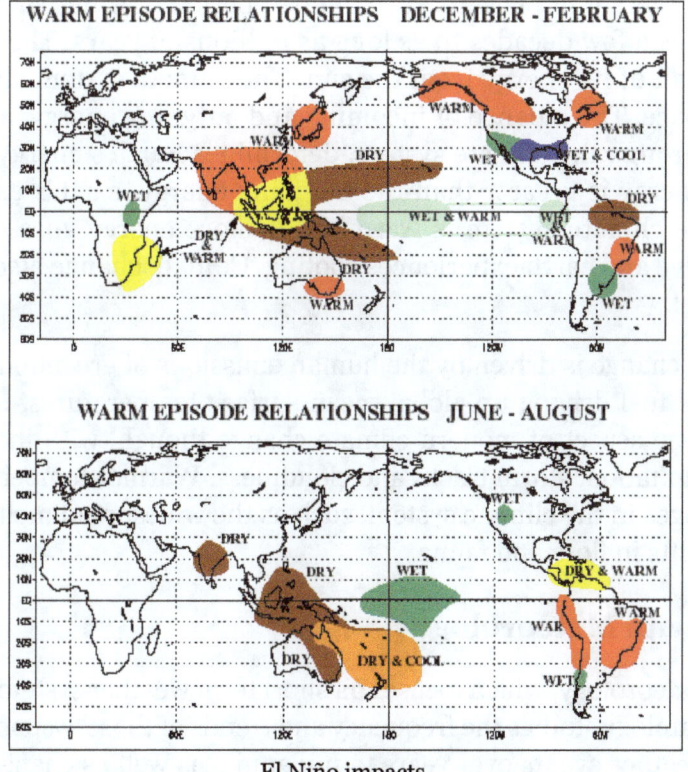

El Niño impacts.

There are different modes of variability: recurring patterns of temperature or other climate variables. They are quantified with different indices. Much in the way the Dow Jones Industrial Average, which is based on the stock prices of 30 companies, is used to represent the fluctuations in the stock market as a whole, climate indices are used to represent the essential elements of climate. Climate indices are generally devised with the twin objectives of simplicity and completeness, and each index typically represents the status and timing of the climate factor it represents. By their very nature, indices are simple, and combine many details into a generalized, overall description of the atmosphere or ocean which can be used to characterize the factors which impact the global climate system.

El Niño–Southern Oscillation (ENSO) is a coupled ocean-atmosphere phenomenon in the Pacific Ocean responsible for most of the global variability in temperature, and has a cycle between two and seven years. The North Atlantic oscillation is a mode of variability that is mainly contained to the lower atmosphere, the troposphere. The layer of atmosphere above, the stratosphere is also capable of creating its own variability, most importantly in the Madden–Julian oscillation (MJO), which has a cycle of approximately 30-60 days. The interdecadal pacific oscillation can create changes in the pacific ocean and lower atmosphere on decadal time scales.

Climatic Change

Climate change occurs when changes in Earth's climate system result in new weather patterns that remain in place for an extended period of time. This length of time can be as short as a few decades to as long as millions of years. The climate system receives nearly all of its energy from the sun. The climate system also gives off energy to outer space. The balance of incoming and outgoing energy, and the passage of the energy through the climate system, determines Earth's energy budget. When the incoming energy is greater than the outgoing energy, earth's energy budget is positive and the climate system is warming. If more energy goes out, the energy budget is negative and earth experiences cooling. Climate change also influences the average sea level.

Modern climate change is driven by the human emissions of greenhouse gas from the burning of fossil fuel driving up global mean surface temperatures. Rising temperatures are only one aspect of modern climate change though, with includes observed changes in precipitation, storm tracks and cloudiness. Warmer temperatures are driving further changes in the climate system, such as the widespread melt of glaciers, sea level rise and shifts in flora and fauna.

Differences with Meteorology

In contrast to meteorology, which focuses on short term weather systems lasting up to a few weeks, climatology studies the frequency and trends of those systems. It studies the periodicity of weather events over years to millennia, as well as changes in long-term average weather patterns, in relation to atmospheric conditions. Climatologists study both the nature of climates – local, regional or global – and the natural or human-induced factors that cause climates to change. Climatology considers the past and can help predict future climate change.

Phenomena of climatological interest include the atmospheric boundary layer, circulation patterns, heat transfer (radiative, convective and latent), interactions between the atmosphere and the oceans and land surface (particularly vegetation, land use and topography), and the chemical and physical composition of the atmosphere.

Use in Weather Forecasting

A more complicated way of making a forecast, the analog technique requires remembering a previous weather event which is expected to be mimicked by an upcoming event. What makes it a difficult technique to use is that there is rarely a perfect analog for an event in the future. Some call this type of forecasting pattern recognition, which remains a useful method of observing rainfall over data voids such as oceans with knowledge of how satellite imagery relates to precipitation rates over land, as well as the forecasting of precipitation amounts and distribution in the future. A variation on this theme is used in medium range forecasting, which is known as teleconnections, when systems in other locations are used to help pin down the location of a system within the surrounding regime. One method of using teleconnections are by using climate indices such as ENSO-related phenomena.

COASTAL GEOGRAPHY

Collapsed Ordovician limestone bank showing coastal erosion.

Coastal geography is the study of the constantly changing region between the ocean and the land, incorporating both the physical geography (i.e. coastal geomorphology, geology and oceanography) and the human geography (sociology and history) of the coast. It includes understanding coastal weathering processes, particularly wave action, sediment movement and weather, and the ways in which humans interact with the coast.

Wave Action and Longshore Drift

The waves of different strengths that constantly hit against the shoreline are the primary movers and shapers of the coastline. Despite the simplicity of this process, the differences between waves and the rocks they hit result in hugely varying shapes.

Port Campbell in southern Australia is a high-energy shoreline.

The effect that waves have depends on their strength. Strong waves, also called destructive waves, occur on high-energy beaches and are typical of winter. They reduce the quantity of sediment present on the beach by carrying it out to bars under the sea. Constructive, weak waves are typical of low-energy beaches and occur most during summer. They do the opposite to destructive waves and increase the size of the beach by piling sediment up onto the berm.

One of the most important transport mechanisms results from wave refraction. Since waves rarely break onto a shore at right angles, the upward movement of water onto the beach (swash) occurs at an oblique angle. However, the return of water (backwash) is at right angles to the beach, resulting in the net movement of beach material laterally. This movement is known as beach drift. The endless cycle of swash and backwash and resulting beach drift can be observed on all beaches. This may differ between coasts.

Probably the most important effect is longshore drift (LSD) (also known as Littoral Drift), the process by which sediment is continuously moved along beaches by wave action. LSD occurs because waves hit the shore at an angle, pick up sediment (sand) on the shore and carry it down the beach at an angle (this is called swash). Due to gravity, the water then falls back perpendicular to the beach, dropping its sediment as it loses energy (this is called backwash). The sediment is then picked up by the next wave and pushed slightly further down the beach, resulting in a continual movement of sediment in one direction. This is the reason why long strips of coast are covered in sediment, not just the areas around river mouths, which are the main sources of beach sediment. LSD is reliant on a constant supply of sediment from rivers and if sediment supply is stopped or sediment falls into a submarine canals at any point along a beach, this can lead to bare beaches further along the shore.

LSD helps create many landforms including barrier islands, bay beaches and spits. In general LSD action serves to straighten the coast because the creation of barriers cuts off bays from the sea while sediment usually builds up in bays because the waves there are weaker (due to wave refraction), while sediment is carried away from the exposed headlands. The lack of sediment on headlands removes the protection of waves from them and makes them more vulnerable to weathering while the gathering of sediment in bays (where longshore drift is unable to remove it) protects the bays from further erosion and makes them pleasant recreational beaches.

Rhossili in Wales is a low-energy shoreline.

Atmospheric Processes

- Onshore winds blowing "up" the beach, pick up sand and move it up the beach to form sand dunes.

- Rain hits the shore and erodes rocks, and carries weathered material to the shoreline to form beaches.

- Warm weather can encourage biological processes to occur more rapidly. In tropical areas some plants and animals protect stones from weathering, while other plants and animals actually eat away at the rocks.

- Temperatures that vary from below to above freezing point result in frost weathering, whereas weather more than a few degrees below freezing point creates sea ice.

Biological Processes

In tropical regions in particular, plants and animals not only affect the weathering of rocks but are a source of sediment themselves. The shells and skeletons of many organisms are of calcium carbonate and when this is broken down it forms sediment, limestone and clay.

Physical Processes

The main physical Weathering process on beaches is salt-crystal growth. Wind carries salt spray onto rocks, where it is absorbed into small pores and cracks within the rocks. There the water evaporates and the salt crystallises, creating pressure and often breaking down the rock. In some beaches calcium carbonate is able to bind together other sediments to form beachrock and in warmer areas dunerock. Wind erosion is also a form of erosion, dust and sand is carried around in the air and slowly erodes rock, this happens in a similar way in the sea were the salt and sand is washed up onto the rocks.

Sea Level Changes (Eustatic Change)

The sea level on earth regularly rises and falls due to climatic changes. During cold periods more of the Earth's water is stored as ice in glaciers while during warm periods it is released and sea levels rise to cover more land. Sea levels are currently quite high, while just 18,000 years ago during the Pleistocene ice age they were quite low. Global warming may result in further rises in the future, which presents a risk to coastal cities as most would be flooded by only small rises. As sea levels rise, fjords and rias form. Fjords are flooded glacial valleys and rias are flooded river valleys. Fjords typically have steep rocky sides, while rias have dendritic drainage patterns typical of drainage zones. As tectonic plates move about the Earth they can rise and fall due to changing pressures and the presence of glaciers. If a beach is moving upwards relative to other plates this is known as isostatic change and raised beaches can be formed.

Land Level Changes (Isostatic Change)

This is found in the U.K. as above the line from the Wash to the Severn estuary, the land was covered in ice sheets during the last ice age. The weight of the ice caused northeast Scotland to sink, displacing the southeast and forcing it to rise. As the ice sheets receded the reverse process happened, as the land was released from the weight. At current estimates the southeast is sinking at a rate of about 2 mm per year, with northeast Scotland rising by the same amount.

Coastal Landforms

Spits

Chesil Beach, a tombolo in Dorset, United Kingdom.

If the coast suddenly changes direction, especially around an estuary, spits are likely to form. Long shore drift pushes the sediment along the beach but when it reaches a turn as in the diagram, the long shore drift does not always easily turn with it, especially near an estuary where the outward flow from a river may push sediment away from the coast. The area may also be shielded from wave action, preventing much long shore drift. On the side of the headland receiving weaker waves, shingle and other large sediments will build up under the water where waves are not strong enough to move them along. This provides a good place for smaller sediments to build up to sea level. The sediment, after passing the headland will accumulate on the other side and not continue down the beach, sheltered both by the headland and the shingle.

Slowly over time sediment simply builds on this area, extending the spit outwards, forming a barrier of sand. Once in a while, the wind direction will change and come from the other direction. During this period the sediment will be pushed along in the other direction. The spit will start to grow backwards, forming a 'hook'. After this time the spit will grow again in the original direction. Eventually the spit will not be able to grow any further because it is no longer sufficiently sheltered from erosion by waves, or because the estuary current prevents sediment resting. Usually in the salty but calm waters behind the spit there will form a salt marshland. Spits often form around the breakwater of artificial harbours requiring dredging.

Occasionally, if there is no estuary then it is possible for the spit to grow across to the other side of the bay and form what is called a bar, or barrier. Barriers come in several varieties, but all form in a manner similar to spits. They usually enclose a bay to form a lagoon. They can join two headlands or join a headland to the mainland. When an island is joined to the mainland with a bar or barrier it is known as a tombolo. This usually occurs due to wave refraction, but can also be caused by isostatic change, a change in the level of the land (e.g. Chesil Beach).

OCEANOGRAPHY

Oceanography also known as oceanology, is the study of the physical and biological aspects of the ocean. It is an important Earth science, which covers a wide range of topics, including ecosystem dynamics; ocean currents, waves, and geophysical fluid dynamics; plate tectonics and the geology of the sea floor; and fluxes of various chemical substances and physical properties within the ocean and across its boundaries. These diverse topics reflect multiple disciplines that oceanographers blend to further knowledge of the world ocean and understanding of processes within: astronomy, biology, chemistry, climatology, geography, geology, hydrology, meteorology and physics. Paleoceanography studies the history of the oceans in the geologic past. An oceanographer is a person who studies many matters concerned with oceans including marine geology, physics, chemistry and biology .

Humans first acquired knowledge of the waves and currents of the seas and oceans in pre-historic times. Observations on tides were recorded by Aristotle and Strabo. Early exploration of the oceans was primarily for cartography and mainly limited to its surfaces and of the animals that fishermen brought up in nets, though depth soundings by lead line were taken.

Although Juan Ponce de León in 1513 first identified the Gulf Stream, and the current was well known to mariners, Benjamin Franklin made the first scientific study of it and gave it its name. Franklin measured water temperatures during several Atlantic crossings and correctly explained the Gulf Stream's cause. Franklin and Timothy Folger printed the first map of the Gulf Stream in 1769–1770.

Information on the currents of the Pacific Ocean was gathered by explorers of the late 18th century, including James Cook and Louis Antoine de Bougainville. James Rennell wrote the first scientific textbooks on oceanography, detailing the current flows of the Atlantic and Indian oceans. During a voyage around the Cape of Good Hope in 1777, he mapped "the banks and currents at the Lagullas". He was also the first to understand the nature of the intermittent current near the Isles of Scilly, (now known as Rennell's Current).

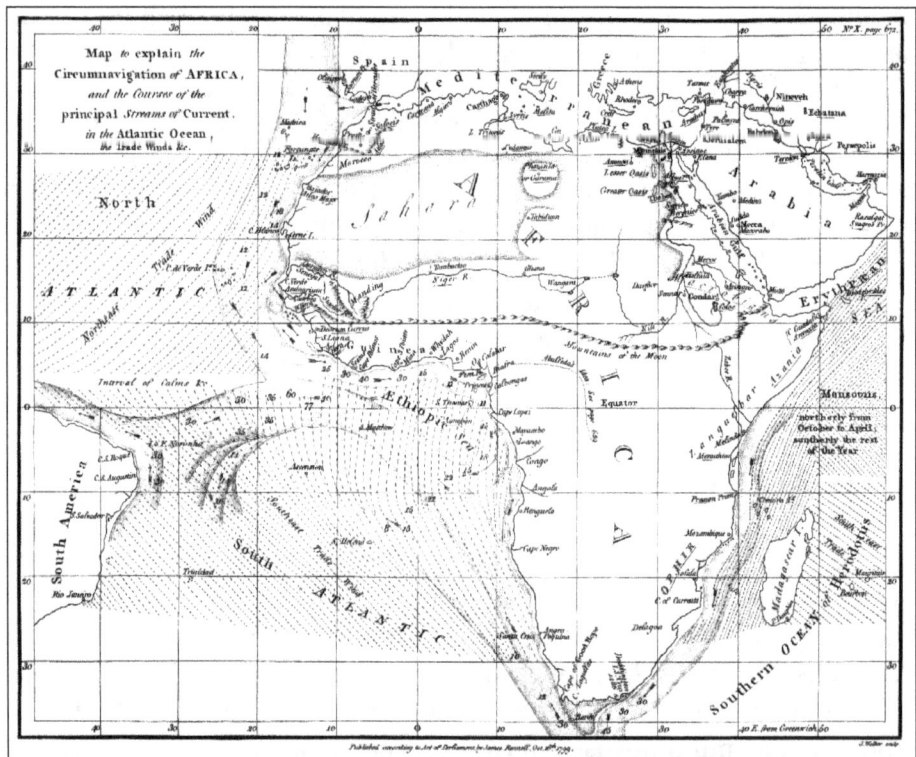

1799 map of the currents in the Atlantic and Indian Oceans.

Sir James Clark Ross took the first modern sounding in deep sea in 1840, and Charles Darwin published a paper on reefs and the formation of atolls as a result of the second voyage of HMS Beagle in 1831–1836. Robert FitzRoy published a four-volume report of

Beagle's three voyages. In 1841–1842 Edward Forbes undertook dredging in the Aegean Sea that founded marine ecology.

The first superintendent of the United States Naval Observatory, Matthew Fontaine Maury devoted his time to the study of marine meteorology, navigation, and charting prevailing winds and currents. His 1855 textbook Physical Geography of the Sea was one of the first comprehensive oceanography studies. Many nations sent oceanographic observations to Maury at the Naval Observatory, where he and his colleagues evaluated the information and distributed the results worldwide.

Modern Oceanography

Despite all this, human knowledge of the oceans remained confined to the topmost few fathoms of the water and a small amount of the bottom, mainly in shallow areas. Almost nothing was known of the ocean depths. The British Royal Navy's efforts to chart all of the world's coastlines in the mid-19th century reinforced the vague idea that most of the ocean was very deep, although little more was known. As exploration ignited both popular and scientific interest in the polar regions and Africa, so too did the mysteries of the unexplored oceans.

HMS Challenger undertook the first global marine research expedition in 1872.

The seminal event in the founding of the modern science of oceanography was the 1872–1876 Challenger expedition. As the first true oceanographic cruise, this expedition laid the groundwork for an entire academic and research discipline. In response to a recommendation from the Royal Society, the British Government announced in 1871 an expedition to explore world's oceans and conduct appropriate scientific investigation. Charles Wyville Thompson and Sir John Murray launched the Challenger expedition. Challenger, leased from the Royal Navy, was modified for scientific work and equipped with separate laboratories for natural history and chemistry. Under the scientific supervision of Thomson, Challenger travelled nearly 70,000 nautical miles (130,000 km) surveying and exploring. On her journey circumnavigating the globe,

492 deep sea soundings, 133 bottom dredges, 151 open water trawls and 263 serial water temperature observations were taken.

Around 4,700 new species of marine life were discovered. The result was the Report Of The Scientific Results of the Exploring Voyage of H.M.S. Challenger during the years 1873–76. Murray, who supervised the publication, described the report as "the greatest advance in the knowledge of our planet since the celebrated discoveries of the fifteenth and sixteenth centuries". He went on to found the academic discipline of oceanography at the University of Edinburgh, which remained the centre for oceanographic research well into the 20th century. Murray was the first to study marine trenches and in particular the Mid-Atlantic Ridge, and map the sedimentary deposits in the oceans. He tried to map out the world's ocean currents based on salinity and temperature observations, and was the first to correctly understand the nature of coral reef development.

In the late 19th century, other Western nations also sent out scientific expeditions (as did private individuals and institutions). The first purpose built oceanographic ship, Albatros, was built in 1882. In 1893, Fridtjof Nansen allowed his ship, Fram, to be frozen in the Arctic ice. This enabled him to obtain oceanographic, meteorological and astronomical data at a stationary spot over an extended period.

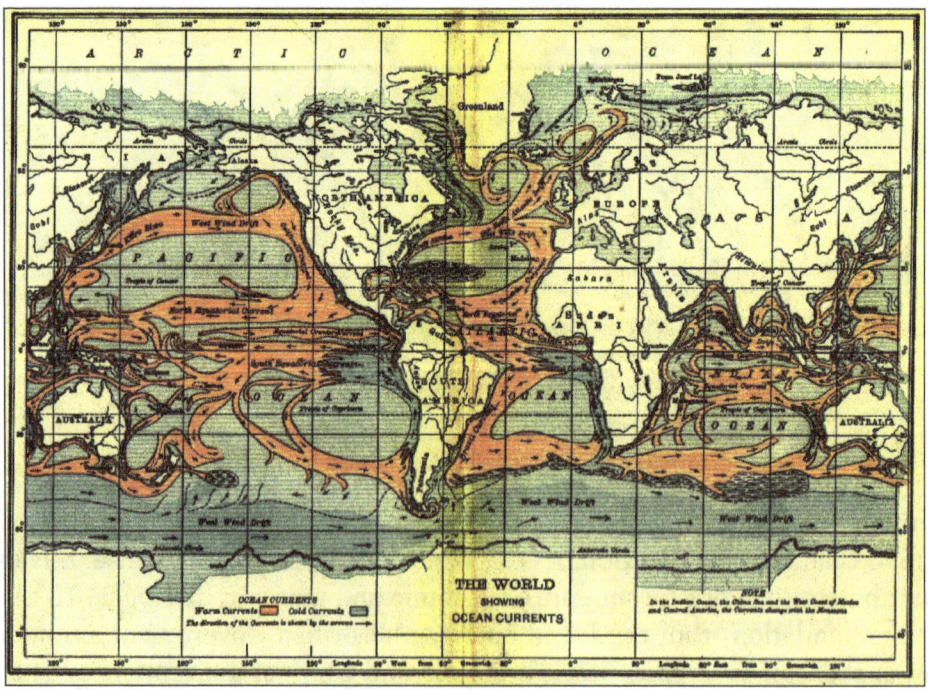

Ocean currents.

In 1881 the geographer John Francon Williams published a seminal book, Geography of the Oceans. Between 1907 and 1911 Otto Krümmel published the Handbuch der Ozeanographie, which became influential in awakening public interest in oceanography. The four-month 1910 North Atlantic expedition headed by John Murray and Johan Hjort was the most ambitious research oceanographic and marine zoological project ever mounted until then, and led to the classic 1912 book The Depths of the Ocean.

The first acoustic measurement of sea depth was made in 1914. Between 1925 and 1927 the "Meteor" expedition gathered 70,000 ocean depth measurements using an echo sounder, surveying the Mid-Atlantic Ridge.

Sverdrup, Johnson and Fleming published The Oceans in 1942, which was a major landmark. The Sea edited by M.N. Hill was published in 1962, while Rhodes Fairbridge's Encyclopedia of Oceanography was published in 1966.

The Great Global Rift, running along the Mid Atlantic Ridge, was discovered by Maurice Ewing and Bruce Heezen in 1953; in 1954 a mountain range under the Arctic Ocean was found by the Arctic Institute of the USSR. The theory of seafloor spreading was developed in 1960 by Harry Hammond Hess. The Ocean Drilling Program started in 1966. Deep-sea vents were discovered in 1977 by Jack Corliss and Robert Ballard in the submersible DSV Alvin.

In the 1950s, Auguste Piccard invented the bathyscaphe and used the bathyscaphe Trieste to investigate the ocean's depths. The United States nuclear submarine Nautilus made the first journey under the ice to the North Pole in 1958. In 1962 the FLIP (Floating Instrument Platform), a 355-foot (108 m) spar buoy, was first deployed.

From the 1970s, there has been much emphasis on the application of large scale computers to oceanography to allow numerical predictions of ocean conditions and as a part of overall environmental change prediction. An oceanographic buoy array was established in the Pacific to allow prediction of El Niño events.

1990 saw the start of the World Ocean Circulation Experiment (WOCE) which continued until 2002.

In recent years studies advanced particular knowledge on ocean acidification, ocean heat content, ocean currents, the El Niño phenomenon, mapping of methane hydrate deposits, the carbon cycle, coastal erosion, weathering and climate feedbacks in regards to climate change interactions.

Study of the oceans is linked to understanding global climate changes, potential global warming and related biosphere concerns. The atmosphere and ocean are linked because of evaporation and precipitation as well as thermal flux (and solar insolation). Wind stress is a major driver of ocean currents while the ocean is a sink for atmospheric carbon dioxide. All these factors relate to the ocean's biogeochemical setup.

Further understanding of the worlds oceans permit scientists to better decide weather changes which in addition guides to a more reliable utilization of earths resources.

Branches

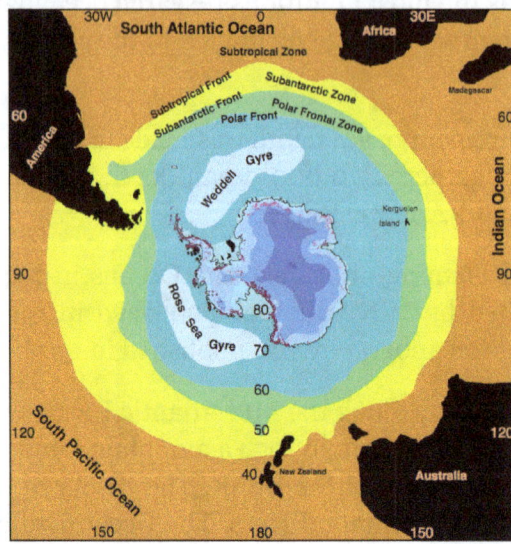

Oceanographic frontal systems on the Southern Hemisphere.

The study of oceanography is divided into these four branches:

Biological Oceanography

Biological oceanography investigates the ecology of marine organisms in the context of the physical, chemical and geological characteristics of their ocean environment and the biology of individual marine organisms.

Chemical Oceanography

Chemical oceanography is the study of the chemistry of the ocean. Whereas chemical oceanography is primarily occupied with the study and understanding of seawater properties and its changes, ocean chemistry focuses primarily on the geochemical cycles. The following is a central topic investigated by chemical oceanography.

Ocean Acidification

Ocean acidification describes the decrease in ocean pH that is caused by anthropogenic carbon dioxide (CO_2) emissions into the atmosphere. Seawater is slightly alkaline and had a preindustrial pH of about 8.2. More recently, anthropogenic activities have steadily increased the carbon dioxide content of the atmosphere; about 30–40% of the added CO_2 is absorbed by the oceans, forming carbonic acid and lowering the pH (now below 8.1) through ocean acidification. The pH is expected to reach 7.7 by the year 2100.

An important element for the skeletons of marine animals is calcium, but calcium carbonate becomes more soluble with pressure, so carbonate shells and skeletons dissolve below the carbonate compensation depth. Calcium carbonate becomes more soluble at lower pH, so ocean acidification is likely to affect marine organisms with calcareous shells, such as oysters, clams, sea urchins and corals, and the carbonate compensation depth will rise closer to the sea surface. Affected planktonic organisms will include pteropods, coccolithophorids and foraminifera, all important in the food chain. In tropical regions, corals are likely to be severely affected as they become less able to build their calcium carbonate skeletons, in turn adversely impacting other reef dwellers.

The current rate of ocean chemistry change seems to be unprecedented in Earth's geological history, making it unclear how well marine ecosystems will adapt to the shifting conditions of the near future. Of particular concern is the manner in which the combination of acidification with the expected additional stressors of higher temperatures and lower oxygen levels will impact the seas.

Geological Oceanography

Geological oceanography is the study of the geology of the ocean floor including plate tectonics and paleoceanography.

Physical Oceanography

Physical oceanography studies the ocean's physical attributes including temperature-salinity structure, mixing, surface waves, internal waves, surface tides, internal tides, and currents. The following are central topics investigated by physical oceanography.

Ocean Currents

Since the early ocean expeditions in oceanography, a major interest was the study of the ocean currents and temperature measurements. The tides, the Coriolis effect, changes in direction and strength of wind, salinity and temperature are the main factors determining ocean currents. The thermohaline circulation (THC) (thermo- referring to temperature and -haline referring to salt content) connects the ocean basins and is primarily dependent on the density of sea water. It is becoming more common to refer to this system as the 'meridional overturning circulation' because it more accurately accounts for other driving factors beyond temperature and salinity.

Examples of sustained currents are the Gulf Stream and the Kuroshio Current which are wind-driven western boundary currents.

Ocean Heat Content

Oceanic heat content (OHC) refers to the heat stored in the ocean. The changes in the ocean heat play an important role in sea level rise, because of thermal expansion. Ocean warming accounts for 90% of the energy accumulation from global warming between 1971 and 2010.

Paleoceanography

Paleoceanography is the study of the history of the oceans in the geologic past with regard to circulation, chemistry, biology, geology and patterns of sedimentation and biological productivity. Paleoceanographic studies using environment models and different proxies enable the scientific community to assess the role of the oceanic processes in the global climate by the reconstruction of past climate at various intervals. Paleoceanographic research is also intimately tied to palaeoclimatology.

QUANTITATIVE REVOLUTION

The quantitative revolution (QR) was a paradigm shift that sought to develop a more rigorous and systematic methodology for the discipline of geography. It came as a response to the inadequacy of regional geography to explain general spatial dynamics. The main claim for the quantitative revolution is that it led to a shift from a descriptive (idiographic) geography to an empirical law-making (nomothetic) geography. The quantitative revolution occurred during the 1950s and 1960s and marked a rapid change in the method behind geographical research, from regional geography into a spatial science.

In the history of geography, the quantitative revolution was one of the four major turning-points of modern geography – the other three being environmental determinism, regional geography and critical geography).

The quantitative revolution had occurred earlier in economics and psychology and contemporaneously in political science and other social sciences and to a lesser extent in history.

Antecedents

During the late 1940s and early 1950s:

- The closing of many geography departments and courses in universities took place, e.g. the abolition of the geography program at Harvard University (a highly prestigious institution) in 1948.

- There was continuing division between human and physical geography – general talk of human geography becoming an autonomous subject.

- Geography was regarded as overly descriptive and unscientific – it was claimed that there was no explanation of why processes or phenomena occurred.

- Geography was seen as exclusively educational – there were few if any applications of contemporary geography.

- Continuing debates regarding what geography is – science, art, humanity or social science – took place.

- After World War II, technology became increasingly important in society, and as a result, nomothetic-based sciences gained popularity and prominence.

All of these events presented a threat to geography's position as an academic subject, and thus geographers began seeking new methods to counter critique.

The Revolution

The quantitative revolution responded to the regional geography paradigm that was dominant at the time. Debates raged predominantly (although not exclusively) in the U.S., where regional geography was the major philosophical school. In the early 1950s, there was a growing sense that the existing paradigm for geographical research was not adequate in explaining how physical, economic, social, and political processes are spatially organized, ecologically related, or how outcomes generated by them are evidence for a given time and place. A growing number of geographers started to express their dissatisfaction with the traditional paradigm of the discipline and its focus on regional geography, deeming the work as too descriptive, fragmented, and non-generalizable. To address these concerns, early critics such as Ackerman suggested the systematization of the discipline. Soon thereafter, a series of debates regarding methodological approaches in geography took place. One of the first illustrations of this was the Schaefer vs. Hartshorne debate. In 1953 Exceptionalism in geography: A Methodological Examination was published. In this work, Schaefer rejected Hartshorne's exceptionalist interpretations about the discipline of geography and having the region as its central object of study. Instead, Schaefer envisioned as the discipline's main objective the establishment of morphological laws through scientific inquiry, i.e. incorporating laws and methods from other disciplines in the social sciences that place a greater emphasis on processes. Hartshorne, on the other hand, addressed Schaefer's criticism in a series of publications, where he dismissed Schaefer's views as subjective and contradictory. He also stressed the importance of describing and classifying places and phenomena, yet admitted that there was room for employing laws of generic relationships in order to maximize scientific understanding. In his view, however, there should be no hierarchy between these two approaches.

While debates about methods carried on, the institutionalization of systematic geography was taking place in the U.S. academy. The geography programs at the University of Iowa, Wisconsin, and Washington were pioneering programs in that respect. At the University of Iowa, Harold McCarty led efforts to establish laws of association between geographical patterns. At the University of Wisconsin, Arthur H. Robinson led efforts to develop statistical methods for map comparison. And at the University of Washington, Edward Ullman and William Garrison worked on developing the field of economic and urban geography, and central place theory. These institutions engendered a generation of geographers that established spatial analysis as part of the research agenda at other institutions including University of Chicago, Northwestern University, Loyola University, The Ohio State University, the University of Michigan, among others.

The changes introduced during the 1950s and 1960s under the banner of bringing 'scientific thinking' to geography led to an increased use of technique-based practices, including an array of mathematical techniques and computerized statistics that improved precision, and theory-based practices to conceptualize location and space in geographical research.

Some of the techniques that epitomize the quantitative revolution include:

- Descriptive statistics;

- Inferential statistics;

- Basic mathematical equations and models, such as gravity model of social physics, or the Coulomb equation;

- Stochastic models using concepts of probability, such as spatial diffusion processes;

- Deterministic models, e.g. Von Thünen's and Weber's location models.

The common factor, linking the above techniques, was a preference for numbers over words and a belief that numerical work had a superior scientific pedigree. Ron JohnstonRon Johnston (geographer) and colleagues at the University of Bristol have published a history of the revolution that stresses changes in substantive focus and philosophical underpinnings as well as methods.

Epistemological Underpinnings

The new method of inquiry led to the development of generalizations about spatial aspects in a wide range of natural and cultural settings. Generalizations may take the form of tested hypotheses, models, or theories, and the research is judged on its scientific validity, turning geography into a nomothetic science.

One of the most significant works to provide a legitimate theoretical and philosophical foundation for the reorientation of geography into a spatial science was David Harvey's book, Explanation in Geography, published in 1969. In this work, Harvey laid out two possible methodologies to explain geographical phenomena: an inductive route where generalizations are made from observation; and a deductive one where, through empirical observation, testable models and hypothesis are formulated and later verified to become scientific laws. He placed preference on the latter method. This positivist approach was countered by critical rationalism, a philosophy advanced by Karl Popper who rejected the idea of verification and maintained that hypothesis can only be falsified. Both epistemological philosophies, however, sought to achieve the same objective: to produce scientific laws and theories.

The paradigm shift had its strongest repercussions in the sub-field of economic and urban geography, especially as it pertains to location theory. However, some geographers–such as Ian Burton–expressed their dissatisfaction with quantification while others – such as Emrys Jones, Peter Lewis, and Golledge and Amedeo – debated the feasibility of law-making. Others, such as F. Luckermann, criticized the scientific explanations offered in geography as conjectural and lacking empirical basis. As a result, even models that were tested failed to accurately depict reality.

By the mid-1960s the quantitative revolution had successfully displaced regional geography from its dominant position and the paradigm shift was evident by the myriad of publications in geographical academic journals and geography textbooks. The adoption of the new paradigm allowed the discipline to be more serviceable to the public and private sectors.

Post-revolution Geography

The quantitative revolution had enormous implications in shaping the discipline of geography into what it looks like today given that its effects led to the spread of positivist (post-positivist) thinking and counter-positivist responses.

The rising interest in the study of distance as a critical factor in understanding the spatial arrangement of phenomena during the revolution led to formulation of the first law of geography by Waldo Tobler. The development of spatial analysis in geography led to more applications in planning process and the further development of theoretical geography offered to geographical research a necessary theoretical background.

The greater use of computers in geography also led to many new developments in geomatics, such as the creation and application of GIS and remote sensing. These new developments allowed geographers for the first time to assess complex models on a full-scale model and over space and time and the relationship between spatial entities. To some extent, the development of geomatics helped obscure the binary between physical and human geography to some extent, as the complexities of the human and natural environments could be assessed on new computable models.

The overwhelming focus on statistical modelling would, eventually, be the undoing of the quantitative revolution. Many geographers became increasingly concerned that these techniques simply put a highly sophisticated technical gloss on an approach to study that was barren of fundamental theory. Other critics argued that it removed the 'human dimension' from a discipline that always prided itself on studying the human and natural world alike. As the 1970s dawned, the quantitative revolution came under direct challenge. The counter-positivist response came as geographers began to expose the inadequacy of quantitative methods to explain and address issues regarding race, gender, class and war. On that regard, David Harvey disregarded earlier works where he advocated for the quantitative revolution and adopted a Marxist theoretical framework. Soon new subfields would emerge in human geography to contribute a new vocabulary for addressing these issues, most notably critical geography and feminist geography. Ron Johnston Ron Johnston (geographer) and Bristol colleagues have argued and documented how quantitative methods can be used in a critical geography. This is a panoramic survey of the legacy of half a century of innovation in spatial science—put into a critical, constructive engagement with half a century of innovation in critical social theory".

PALAEOGEOGRAPHY

Palaeogeography is the ancient geography of Earth's surface. Earth's geography is constantly changing: continents move as a result of plate tectonic interactions; mountain ranges are thrust up and erode; and sea levels rise and fall as the volume of the ocean basins change. These geographic changes can be traced through the study of the rock and fossil record, and data can be used to create palaeogeographic maps, which illustrate how the continents have moved and how the past locations of mountains, lowlands, shallow seas, and deep ocean basins have changed.

The study of paleogeography has two principal goals. The first is to map the past positions of the continents and ocean basins, and the second is to illustrate Earth's changing geographic features through time.

Mapping Past Continents and Oceans

The past positions of the continents can be determined by using six major lines of evidence: paleomagnetism, linear magnetic anomalies, hot-spot tracks, paleobiogeography, paleoclimatology, and geologic and tectonic history.

Paleomagnetism

By measuring the remanent magnetic field often preserved in rocks containing iron-bearing minerals, paleomagnetic analysis can determine whether a rock was magnetized near one of Earth's poles or near the Equator. Iron-bearing minerals forming in igneous rock align themselves with Earth's magnetic field as the molten rock cools. These minerals also align themselves when they are deposited in sediments, and they retain their orientation as they lithify into sedimentary rock. Lines of force in Earth's magnetic field are parallel to the planet's surface at the Equator and are vertical at the poles. Therefore, iron-bearing minerals formed or deposited at low latitudes will be nearly parallel to Earth's surface, while those at high latitudes will dip steeply. If the rocks are later transported by tectonic processes, their original latitude of deposition can be determined by their orientation. Paleomagnetism provides direct evidence of a continent's past north-south (latitudinal) position, but it does not constrain its east-west (longitudinal) position.

Linear Magnetic Anomalies

Earth's magnetic field has another important property. Like the Sun's magnetic field, Earth's magnetic field periodically "flips," or reverses polarity—that is, the North and South poles switch places. Fluctuations, or anomalies in the intensity of the magnetic field, occur at the boundaries between normally magnetized sea floor and sea floor magnetized in the reversed direction. The age of these magnetic anomalies can be established by using fossil evidence and radiometric age determinations. Because these magnetic anomalies form at oceanic ridges, they tend to be long, linear features (hence the name linear magnetic anomalies) that are symmetrically disposed about ridge axes. The past positions of the continents during the last 150 million years (the maximum age of most of the ocean floor) can be directly reconstructed by superimposing linear magnetic anomalies of the same age, in effect "undoing" the results of sea-floor spreading since that time.

Hot-spot Tracks

Some of the world's volcanoes are formed by jets of molten rock that arise at the boundary between Earth's core and mantle (at a depth of about 2,900 km, or 1,800 miles). These rising plumes, or hot spots, puncture the lithosphere, and, as a tectonic plate moves across the hot spot, a line of islands is generated. The island directly above the hot spot is the youngest, and islands become progressively older with distance from the hot spot. There are more than a dozen well-documented hot-spot tracks. Perhaps the most obvious is the Hawaiian Islands, which trace an east-west arc across the central Pacific Ocean. Hot-spot tracks accurately record plate motions and can be used to determine the past latitudinal and longitudinal position of the continents.

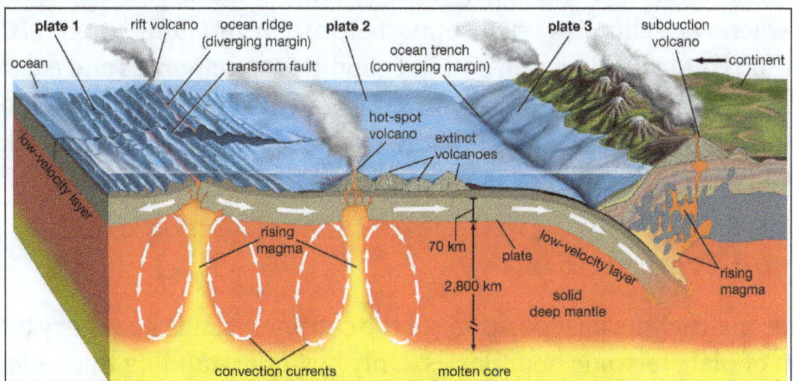

Volcanic activity and the Earth's tectonic plates. Stratovolcanoes tend to form at subduction zones, or convergent plate margins, where an oceanic plate slides beneath a continental plate and contributes to the rise of magma to the surface. At rift zones, or divergent margins, shield volcanoes tend to form as two oceanic plates pull slowly apart and magma effuses upward through the gap. Volcanoes are not generally found at strike-slip zones, where two plates slide laterally past each other. "Hot spot" volcanoes may form where plumes of lava rise from deep within the mantle to the Earth's crust far from any plate margins.

Paleobiogeography

The past distribution of plants and animals can give important clues about the latitudinal position of the continents as well as their relative positions. Cold-water faunas can often be distinguished from warm-water faunas, and ancient floras reflect both paleotemperature and paleorainfall. The diversity of plants and animals tends to increase toward the Equator, and the adaptations of plants (such as smooth-edged leaves in the tropics and serrated-edged leaves in the temperate belts) are often good indicators of the amount of ancient rainfall.

The similarity or dissimilarity of faunas and floras on different continents can also be used to estimate their geographic proximity. In addition, the evolutionary history of groups of plants and animals on different continents can reveal when these continents were connected or isolated from each other. For example, Australia's unique marsupial fauna is the result of its isolation from the other continents at the time when placental mammals were evolving on the other continents during the early Paleogene Period.

Paleoclimatology

Earth's climate is primarily a result of the redistribution of the Sun's energy across the surface of the globe. It is warm near the Equator and cool near the poles. Wetness or rainfall also varies systematically from the Equator to the pole in alternating bands. It is wet near the Equator, dry in the subtropics, wet in the temperate belts, and dry near the poles. Certain kinds of rocks form under specific climatic conditions. For example, coals occur where wet climates once supported lush vegetation; bauxite (the principal ore of aluminum) is formed in warm and wet conditions, evaporites and calcretes require warmth and aridity to form; and tillites are deposited during the movement of glacial ice. The ancient distribution of these and other rock types can indicate how the global climate has changed through time and how the continents have traveled across the climatic belts.

Geologic and Tectonic History

In order to reconstruct the past positions of the continents, it is necessary to understand the evolution of plate tectonic boundaries. Only by understanding the regional geologic and tectonic history of an area can the location and timing of rifting events, subduction activity, continental collision, and other major plate tectonic events be determined.

Mapping Past Geographic Features

Palaeogeographic features include mountain ranges, lowlands, shallow seas, and the deep ocean basins. Some palaeogeographic features change very slowly and are easy to map. Others change very rapidly, so that any palaeogeographic mapping is, at best, an approximation.

Continents and Ocean Basins

The two major palaeogeographic features are the continents and the ocean basins. Since early Precambrian time, Earth has been divided into deep ocean basins (average depth, 3.5 km, or 2.2 miles) and high-standing continents (average elevation, about 800 metres, or 2,600 feet). Continental lithosphere stands high above the ocean basins because it is less dense and is not easily subducted, or recycled back into Earth's interior. Consequently, continents are made up of very old rocks, some dating back over 4 billion years. The amount of continental lithosphere has probably changed very little during the last 2.6 billion years—possibly increasing 10 to 15 percent. What has changed is the shape and the distribution of continents across the globe.

The ocean basins are also ancient palaeogeographic features. Oceanic lithosphere is continuously created at oceanic ridges and then recycled back into Earth's interior at subduction zones.

Mountain Ranges

In contrast to the continents and ocean basins, which are permanent geographic features, the height and location of mountain belts constantly change. Mountain belts form either where oceanic lithosphere is subducted beneath the margin of a continent, giving rise to a linear range of mountains such as the Andes of western South America, or where continents collide, forming high mountains and broad plateaus such as the Himalayas and the Tibetan Plateau of Asia. Less-extensive mountains can also form when continents rift apart (as is happening today in the East African Rift) or where hot spots form volcanic uplifts.

In most cases, mountain ranges take tens of millions of years to form and, depending on the climate, may last for hundreds of millions of years. Though the Appalachian Mountains of the eastern United States were formed more than 300 million years ago, as a result of the collision of North America and western Africa, remnants of this collisional mountain belt still reach heights of over than 2,000 metres (6,600 feet). The Himalayan mountains, the world's tallest mountain range, began to rise from the sea nearly 50 million years ago when northern India collided with Eurasia.

Shorelines and Continental Margins

In contrast to mountain ranges, which take tens to hundreds of millions of years to uplift and erode, the location of Earth's shorelines can change rapidly. The familiar shapes that characterize today's shorelines such as Hudson's Bay, the Florida peninsula, or the numerous fiords of Norway are all less than 12,000 years old. The shape of the modern coastlines is the result of a rise in sea level of 70 metres (230 feet) that took place in the last 12,000 years as the last great ice sheet that covered much of North America and Europe melted.

It is important to note that the shoreline, though the edge of land, is not the edge of the continent. In most cases the continent extends seaward hundreds of kilometres beyond the shoreline. The actual edge of the continent in most cases is marked by the transition from the continental slope to the continental rise. This steep bathymetric gradient marks the boundary between continental and oceanic crust.

Agents of Palaeogeographic Change

The ancient distribution of land and sea, probably the single most important aspect of paleogeography, is a function of both continental topography and sea-level change. Though topography changes slowly (over tens of millions of years), global sea level can change rapidly (over tens of thousand of years). When sea level rises, the continents are flooded and shorelines move landward. Throughout much of Earth's history, sea level was higher than it is today, and vast areas of the continents were flooded by shallow seas.

Several factors can affect sea-level change. One factor is the amount of ice on the continents. At times when the continents are covered by great ice sheets, sea level is low, and the continents are more exposed. The last glacial maximum was 18,000 years ago. Other important global episodes of glaciation occurred 300 million, 450 million, and 650 million years ago. The oldest known glacial episode occurred in the Precambrian, approximately 2.2 billion years ago. For the last 20 million years, the continents and their margins have been largely high and dry because there has been a significant amount of ice on Antarctica and there has been extensive mountain-building in Asia.

Sea level also changes more slowly (over tens of millions of years) owing to changes in the volume of the ocean basins. During the Precambrian, gases escaped from Earth's interior and contributed to the formation of water vapour in the atmosphere. The vapour eventually condensed on the cooling surface to form the world's oceans. However, there has been no significant addition to the volume of water on Earth since early Precambrian times. Changes in sea level, therefore, are due not to changes in the amount of water on Earth but rather to changes in the shape and volume of the ocean basins. Plate tectonics and, in particular, sea-floor spreading control the shape and volume of the ocean basins.

References

- Physical-geography-overview-1435345: thoughtco.com, Retrieved 13 July, 2020

- Importance-physical-geography, geology: qsstudy.com, Retrieved 12 April, 2020

- Garcia-Castellanos, D., 2007. The role of climate during high plateau formation. Insights from numerical experiments. Earth Planet. Sci. Lett. 257, 372-390, doi:10.1016/j.epsl.2007.02.039

- Browne, Janet (1983). The secular ark: studies in the history of biogeography. New Haven: Yale University Press. ISBN 978-0-300-02460-9

- Cohen, A.; Holcomb, M. (2009). "Why Corals Care About Ocean Acidification: Uncovering the Mechanism" (PDF). Oceanography. 24 (4): 118–127. doi:10.5670/oceanog.2009.102. hdl:1912/3179. Archived from the original (PDF) on 6 November 2013

- Plain, encyclopedia: nationalgeographic.org, Retrieved 19 June, 2020

- Paleogeography, science: britannica.com, Retrieved 13 May, 2020

- Quantitative methods I: The world we have lost – or where we started from;https://journals.sagepub.com/doi/abs/10.1177/0309132518774967 ; Quantitative methods II: How we moved on – Decades of change in philosophy, focus and methods; https://journals.sagepub.com/doi/abs/10.1177/0309132519869451

4

Tools and Techniques used in Geography

There are various tools and techniques that are used in geography. It includes the use of geostatistics, cartography, geographic information system, geographic coordinate system, satellite imagery, geoinformatics, etc. This chapter closely examines these tools and techniques used in geography for a thorough understanding of the subject.

CARTOGRAPHY

Cartography is the art and science of graphically representing a geographical area, usually on a flat surface such as a map or chart. It may involve the superimposition of political, cultural, or other nongeographical divisions onto the representation of a geographical area.

Ptolemy's map of the world.

Cartography is an ancient discipline that dates from the prehistoric depiction of hunting and fishing territories. The Babylonians mapped the world in a flattened, disk-shaped form, but Claudius Ptolemaeus (Ptolemy) established the basis for subsequent efforts in the 2nd century ce with his eight-volume work Geōgraphikē hyphēgēsis (Guide to Geography) that showed a spherical Earth. Maps produced during the Middle Ages followed Ptolemy's guide, but they used Jerusalem as the central feature and placed East at the top. Those representations are often called T-maps because they show only three continents (Europe, Asia, and Africa), separated by the "T" formed by the Mediterranean Sea and the Nile River. More accurate geographical representation began in the 14th century when portolan (seamen's) charts were compiled for navigation.

The discovery of the New World by Europeans led to the need for new techniques in cartography, particularly for the systematic representation on a flat surface of the features of a curved surface—generally referred to as a projection (e.g., Mercator projection, cylindrical projection, and Lambert conformal projection). During the 17th and 18th centuries there was a vast outpouring of printed maps of ever-increasing accuracy and sophistication. Systematic surveys were undertaken involving triangulation that greatly improved map reliability and precision. Noteworthy among the scientific methods introduced later was the use of the telescope for determining the length of a degree of longitude.

Modern cartography largely involves the use of aerial and, increasingly, satellite photographs as a base for any desired map or chart. The procedures for translating photographic data into maps are governed by the principles of photogrammetry and yield a degree of accuracy previously unattainable. The remarkable improvements in satellite photography since the late 20th century and the general availability on the Internet of satellite images have made possible the creation of Google Earth and other databases that are widely available online. Satellite photography has also been used to create highly detailed maps of features of the Moon and of several planets in our solar system and their satellites. In addition, the use of geographic information systems (GIS) has been indispensible in expanding the scope of cartographic subjects.

GEOGRAPHIC INFORMATION SYSTEM

A Geographic Information System (GIS) is a system designed to capture, store, manipulate, analyze, manage, and present spatial or geographic data. GIS applications are tools that allow users to create interactive queries (user-created searches), analyze spatial information, edit data in maps, and present the results of all these operations. GIS (more commonly GIScience) sometimes refers to geographic information science (GIScience), the science underlying geographic concepts, applications, and systems. Since the mid-1980s, geographic information systems have become valuable tool used to support a variety of city and regional planning functions.

GIS can refer to a number of different technologies, processes, techniques and methods. It is attached to many operations and has many applications related to engineering, planning, management, transport/logistics, insurance, telecommunications, and business. For that reason, GIS and location intelligence applications can be the foundation for many location-enabled services that rely on analysis and visualization.

GIS can relate unrelated information by using location as the key index variable. Locations or extents in the Earth space–time may be recorded as dates/times of occurrence, and x, y, and z coordinates representing, longitude, latitude, and elevation, respectively. All Earth-based spatial–temporal location and extent references should be relatable to one another and ultimately to a "real" physical location or extent. This key characteristic of GIS has begun to open new avenues of scientific inquiry.

History of Development

The first known use of the term "geographic information system" was by Roger Tomlinson in the year 1968 in his paper "A Geographic Information System for Regional Planning". Tomlinson is also acknowledged as the "father of GIS".

E. W. Gilbert's version of 1855 map of the Soho cholera outbreak showing the clusters of cholera cases in the London epidemic of 1854.

Previously, one of the first applications of spatial analysis in epidemiology is the 1832 "Rapport sur la marche et les effets du choléra dans Paris et le département de la Seine". The French geographer Charles Picquet represented the 48 districts of the city of Paris by halftone color gradient according to the number of deaths by cholera per 1,000 inhabitants. In 1854 John Snow determined the source of a cholera outbreak in London by marking points on a map depicting where the cholera victims lived, and connecting the cluster that he found with a nearby water source. This was one of the earliest successful uses of a geographic methodology in epidemiology. While the basic elements of topography and theme existed previously in cartography, the John Snow map was unique, using cartographic methods not only to depict but also to analyze clusters of geographically dependent phenomena.

The early 20th century saw the development of photozincography, which allowed maps to be split into layers, for example one layer for vegetation and another for water. This was particularly used for printing contours – drawing these was a labour-intensive task but having them on a separate layer meant they could be worked on without the other layers to confuse the draughtsman. This work was originally drawn on glass plates but later plastic film was introduced, with the advantages of being lighter, using less storage space and being less brittle, among others. When all the layers were finished, they were combined into one image using a large process camera. Once color printing came in, the layers idea was also used for creating separate printing plates for each color. While the use of layers much later became one of the main typical features of a contemporary GIS, the photographic process just described is not considered to be a GIS in itself – as the maps were just images with no database to link them to.

Two additional developments are notable in the early days of GIS: Ian McHarg's publication "Design with Nature" and its map overlay method and the introduction of a street network into the U.S. Census Bureau's DIME (Dual Independent Map Encoding) system.

Computer hardware development spurred by nuclear weapon research led to general-purpose computer "mapping" applications by the early 1960s.

The year 1960 saw the development of the world's first true operational GIS in Ottawa, Ontario, Canada, by the federal Department of Forestry and Rural Development. Developed by Dr. Roger Tomlinson, it was called the Canada Geographic Information System (CGIS) and was used to store, analyze, and manipulate data collected for the Canada Land Inventory – an effort to determine the land capability for rural Canada by mapping information about soils, agriculture, recreation, wildlife, waterfowl, forestry and land use at a scale of 1:50,000. A rating classification factor was also added to permit analysis.

CGIS was an improvement over "computer mapping" applications as it provided capabilities for overlay, measurement, and digitizing/scanning. It supported a national coordinate system that spanned the continent, coded lines as arcs having a true embedded topology and it stored the attribute and locational information in separate files. As a result of this, Tomlinson has become known as the "father of GIS", particularly for his use of overlays in promoting the spatial analysis of convergent geographic data.

CGIS lasted into the 1990s and built a large digital land resource database in Canada. It was developed as a mainframe-based system in support of federal and provincial resource planning and management. Its strength was continent-wide analysis of complex datasets. The CGIS was never available commercially.

In 1964 Howard T. Fisher formed the Laboratory for Computer Graphics and Spatial Analysis at the Harvard Graduate School of Design, where a number of important theoretical concepts in spatial data handling were developed, and which by the 1970s had distributed seminal software code and systems, such as SYMAP, GRID, and ODYSSEY – that served as sources for subsequent commercial development—to universities, research centers and corporations worldwide.

By the late 1970s two public domain GIS systems (MOSS and GRASS GIS) were in development, and by the early 1980s, M&S Computing (later Intergraph) along with Bentley Systems Incorporated for the CAD platform, Environmental Systems Research Institute (ESRI), CARIS (Computer Aided Resource Information System), MapInfo Corporation and ERDAS (Earth Resource Data Analysis System) emerged as commercial vendors of GIS software, successfully incorporating many of the CGIS features, combining the first generation approach to separation of spatial and attribute information with a second generation approach to organizing attribute data into database structures.

In 1986, Mapping Display and Analysis System (MIDAS), the first desktop GIS product was released for the DOS operating system. This was renamed in 1990 to MapInfo for Windows when it was ported to the Microsoft Windows platform. This began the process of moving GIS from the research department into the business environment.

By the end of the 20th century, the rapid growth in various systems had been consolidated and standardized on relatively few platforms and users were beginning to explore viewing GIS data over the Internet, requiring data format and transfer standards. More recently, a growing number of free, open-source GIS packages run on a range of operating systems and can be customized to perform specific tasks. Increasingly geospatial data and mapping applications are being made available via the World Wide Web.

Techniques and Technology

Modern GIS technologies use digital information, for which various digitized data creation methods are used. The most common method of data creation is digitization, where a hard copy map or survey plan is transferred into a digital medium through the use of a CAD program, and geo-referencing capabilities. With the wide availability of ortho-rectified imagery (from satellites, aircraft, Helikites and UAVs), heads-up digitizing is becoming the main avenue through which geographic data is extracted. Heads-up digitizing involves the tracing of geographic data directly on top of the aerial imagery instead of by the traditional method of tracing the geographic form on a separate digitizing tablet (heads-down digitizing).

Geoprocessing is a GIS operation used to manipulate spatial data. A typical geoprocessing operation takes an input dataset, performs an operation on that dataset, and returns the result of the operation as an output dataset. Common geoprocessing operations include geographic feature overlay, feature selection and analysis, topology processing, raster processing, and data conversion. Geoprocessing allows for definition, management, and analysis of information used to form decisions.

Relating Information from Different Sources

GIS uses spatio-temporal (space-time) location as the key index variable for all other information. Just as a relational database containing text or numbers can relate many different tables using common key index variables, GIS can relate otherwise unrelated information by using location as the key index variable. The key is the location and extent in space-time.

Any variable that can be located spatially, and increasingly also temporally, can be referenced using a GIS. Locations or extents in Earth space–time may be recorded as dates/times of occurrence, and x, y, and z coordinates representing, longitude, latitude, and elevation, respectively. These GIS coordinates may represent other quantified systems of temporo-spatial reference (for example, film frame number, stream gage station, highway mile-marker, surveyor benchmark, building address, street intersection, entrance gate, water depth sounding, POS or CAD drawing origin/units). Units applied to recorded temporal-spatial data can vary widely (even when using exactly the same data, see map projections), but all Earth-based spatial–temporal location and extent references should, ideally, be relatable to one another and ultimately to a "real" physical location or extent in space–time.

Related by accurate spatial information, an incredible variety of real-world and projected past or future data can be analyzed, interpreted and represented. This key characteristic of GIS has begun to open new avenues of scientific inquiry into behaviors and patterns of real-world information that previously had not been systematically correlated.

GIS Uncertainties

GIS accuracy depends upon source data, and how it is encoded to be data referenced. Land surveyors have been able to provide a high level of positional accuracy utilizing the GPS-derived positions. High-resolution digital terrain and aerial imagery, powerful computers and Web technology are changing the quality, utility, and expectations of GIS to serve society on a grand scale, but nevertheless there are other source data that affect overall GIS accuracy like paper maps, though these may be of limited use in achieving the desired accuracy.

In developing a digital topographic database for a GIS, topographical maps are the main source, and aerial photography and satellite imagery are extra sources for collecting data and identifying attributes which can be mapped in layers over a location facsimile of scale. The scale of a map and geographical rendering area representation type are very important aspects since the information content depends mainly on the scale set and resulting locatability of the map's representations. In order to digitize a map, the map has to be checked within theoretical dimensions, then scanned into a raster format, and resulting raster data has to be given a theoretical dimension by a rubber sheeting/warping technology process.

A quantitative analysis of maps brings accuracy issues into focus. The electronic and other equipment used to make measurements for GIS is far more precise than the machines of conventional map analysis. All geographical data are inherently inaccurate, and these inaccuracies will propagate through GIS operations in ways that are difficult to predict.

Data Representation

GIS data represents real objects (such as roads, land use, elevation, trees, waterways, etc.) with digital data determining the mix. Real objects can be divided into two abstractions: discrete objects (e.g., a house) and continuous fields (such as rainfall amount, or elevations). Traditionally, there are two broad methods used to store data in a GIS for both kinds of abstractions mapping references: raster images and vector. Points, lines, and polygons are the stuff of mapped location attribute references. A new hybrid method of storing data is that of identifying point clouds, which combine three-dimensional points with RGB information at each point, returning a "3D color image". GIS thematic maps then are becoming more and more realistically visually descriptive of what they set out to show or determine.

Data Capture

Example of hardware for mapping (GPS and laser rangefinder) and data collection (rugged computer). The current trend for geographical information system (GIS) is that accurate mapping and data analysis are completed while in the field. Depicted hardware (field-map technology) is used mainly for forest inventories, monitoring and mapping.

Data capture—entering information into the system—consumes much of the time of GIS practitioners. There are a variety of methods used to enter data into a GIS where it is stored in a digital format.

Existing data printed on paper or PET film maps can be digitized or scanned to produce digital data. A digitizer produces vector data as an operator traces points, lines, and polygon boundaries from a map. Scanning a map results in raster data that could be further processed to produce vector data.

Survey data can be directly entered into a GIS from digital data collection systems on survey instruments using a technique called coordinate geometry (COGO). Positions from a global navigation satellite system (GNSS) like Global Positioning System can also be collected and then imported into a GIS. A current trend in data collection gives users the ability to utilize field computers with the ability to edit live data using wireless connections or disconnected editing sessions. This has been enhanced by the availability of low-cost mapping-grade GPS units with decimeter accuracy in real time. This eliminates the need to post process, import, and update the data in the office after fieldwork has been collected. This includes the ability to incorporate positions collected using a laser rangefinder. New technologies also allow users to create maps as well as analysis directly in the field, making projects more efficient and mapping more accurate.

Remotely sensed data also plays an important role in data collection and consist of sensors attached to a platform. Sensors include cameras, digital scanners and lidar, while platforms usually consist of aircraft and satellites. In England in the mid 1990s, hybrid kite/balloons called helikites first pioneered the use of compact airborne digital cameras as airborne geo-information systems. Aircraft measurement software, accurate to 0.4 mm was used to link the photographs and measure the ground. Helikites are inexpensive and gather more accurate data than aircraft. Helikites can be used over roads, railways and towns where unmanned aerial vehicles (UAVs) are banned.

Recently aerial data collection is becoming possible with miniature UAVs. For example, the Aeryon Scout was used to map a 50-acre area with a ground sample distance of 1 inch (2.54 cm) in only 12 minutes.

The majority of digital data currently comes from photo interpretation of aerial photographs. Soft-copy workstations are used to digitize features directly from stereo pairs of digital photographs. These systems allow data to be captured in two and three dimensions, with elevations measured directly from a stereo pair using principles of photogrammetry. Analog aerial photos must be scanned before being entered into a soft-copy system, for high-quality digital cameras this step is skipped.

Satellite remote sensing provides another important source of spatial data. Here satellites use different sensor packages to passively measure the reflectance from parts of the electromagnetic spectrum or radio waves that were sent out from an active sensor such as radar. Remote sensing collects raster data that can be further processed using different bands to identify objects and classes of interest, such as land cover.

When data is captured, the user should consider if the data should be captured with either a relative accuracy or absolute accuracy, since this could not only influence how information will be interpreted but also the cost of data capture.

After entering data into a GIS, the data usually requires editing, to remove errors, or further processing. For vector data it must be made "topologically correct" before it can be used for some advanced analysis. For example, in a road network, lines must connect with nodes at an intersection. Errors such as undershoots and overshoots must also be removed. For scanned maps, blemishes on the source map may need to be removed from the resulting raster. For example, a fleck of dirt might connect two lines that should not be connected.

Raster-to-vector Translation

Data restructuring can be performed by a GIS to convert data into different formats. For example, a GIS may be used to convert a satellite image map to a vector structure by generating lines around all cells with the same classification, while determining the cell spatial relationships, such as adjacency or inclusion.

More advanced data processing can occur with image processing, a technique developed in the late 1960s by NASA and the private sector to provide contrast enhancement, false color rendering and a variety of other techniques including use of two dimensional Fourier transforms. Since digital data is collected and stored in various ways, the two data sources may not be entirely compatible. So a GIS must be able to convert geographic data from one structure to another. In so doing, the implicit assumptions behind different ontologies and classifications require analysis. Object ontologies have gained increasing prominence as a consequence of object-oriented programming and sustained work by Barry Smith and co-workers.

Projections, Coordinate Systems and Registration

The earth can be represented by various models, each of which may provide a different set of coordinates (e.g., latitude, longitude, elevation) for any given point on the Earth's surface. The simplest model is to assume the earth is a perfect sphere. As more measurements of the earth have accumulated, the models of the earth have become more sophisticated and more accurate. In fact, there are models called datums that apply to different areas of the earth to provide increased accuracy, like NAD83 for U.S. measurements, and the World Geodetic System for worldwide measurements.

Spatial Analysis with Geographical Information System (GIS)

GIS spatial analysis is a rapidly changing field, and GIS packages are increasingly including analytical tools as standard built-in facilities, as optional toolsets, as add-ins or 'analysts'. In many instances these are provided by the original software suppliers (commercial vendors or collaborative non commercial development teams), while in other cases facilities have been developed and are provided by third parties. Furthermore, many products offer software development kits (SDKs), programming languages and language support, scripting facilities and special interfaces for developing one's own analytical tools or variants. The increased availability has created a new dimension to business intelligence termed "spatial intelligence" which, when openly delivered via intranet, democratizes access to geographic and social network data. Geospatial intelligence, based on GIS spatial analysis, has also become a key element for security. GIS as a whole can be described as conversion to a vectorial representation or to any other digitisation process.

Slope and Aspect

Slope can be defined as the steepness or gradient of a unit of terrain, usually measured as an angle in degrees or as a percentage. Aspect can be defined as the direction in which a unit of terrain faces. Aspect is usually expressed in degrees from north. Slope, aspect, and surface curvature in terrain analysis are all derived from neighborhood operations using elevation values of a cell's adjacent neighbours. Slope is a function of resolution, and the spatial resolution used to calculate slope and aspect should always be specified. Various authors have compared techniques for calculating slope and aspect.

The following method can be used to derive slope and aspect: The elevation at a point or unit of terrain will have perpendicular tangents (slope) passing through the point, in an east-west and north-south direction. These two tangents give two components, $\partial z/\partial x$ and $\partial z/\partial y$, which then be used to determine the overall direction of slope, and the aspect of the slope. The gradient is defined as a vector quantity with components equal to the partial derivatives of the surface in the x and y directions.

The calculation of the overall 3×3 grid slope S and aspect A for methods that determine east-west and north-south component use the following formulas respectively:

$$\text{an } S = \sqrt{\left(\frac{\partial z}{\partial x}\right)^2 + \left(\frac{\partial z}{\partial y}\right)^2}$$

$$\tan A = \frac{\left(\dfrac{-\partial z}{\partial y}\right)}{\left(\dfrac{\partial z}{\partial x}\right)}$$

Zhou and Liu describe another formula for calculating aspect, as follows:

$$A = 270^\circ + \arctan\left(\frac{\left(\dfrac{\partial z}{\partial x}\right)}{\left(\dfrac{\partial z}{\partial y}\right)}\right) - 90^\circ \cdot \frac{\left(\dfrac{\partial z}{\partial y}\right)}{\left|\dfrac{\partial z}{\partial y}\right|}$$

Data Analysis

It is difficult to relate wetlands maps to rainfall amounts recorded at different points such as airports, television stations, and schools. A GIS, however, can be used to depict two- and three-dimensional characteristics of the Earth's surface, subsurface, and atmosphere from information points. For example, a GIS can quickly generate a map with isopleth or contour lines that indicate differing amounts of rainfall. Such a map can be thought of as a rainfall contour map. Many sophisticated methods can estimate the characteristics of surfaces from a limited number of point measurements. A two-dimensional contour map created from the surface modeling of rainfall point measurements may be overlaid and analyzed with any other map in a GIS covering the same area. This GIS derived map can then provide additional information - such as the viability of water power potential as a renewable energy source. Similarly, GIS can be used to compare other renewable energy resources to find the best geographic potential for a region.

Additionally, from a series of three-dimensional points, or digital elevation model, isopleth lines representing elevation contours can be generated, along with slope analysis, shaded relief, and other elevation products. Watersheds can be easily defined for any given reach, by computing all of the areas contiguous and uphill from any given point of interest. Similarly, an expected thalweg of where surface water would want to travel in intermittent and permanent streams can be computed from elevation data in the GIS.

Topological Modeling

A GIS can recognize and analyze the spatial relationships that exist within digitally stored spatial data. These topological relationships allow complex spatial modelling and analysis to be performed. Topological relationships between geometric entities traditionally include adjacency (what adjoins what), containment (what encloses what), and proximity (how close something is to something else).

Geometric Networks

Geometric networks are linear networks of objects that can be used to represent interconnected features, and to perform special spatial analysis on them. A geometric network is composed of edges, which are connected at junction points, similar to graphs in mathematics and computer science. Just like graphs, networks can have weight and flow assigned to its edges, which can be used to represent various interconnected features more accurately. Geometric networks are often used to model road networks and public utility networks, such as electric, gas, and water networks. Network modeling is also commonly employed in transportation planning, hydrology modeling, and infrastructure modeling.

Hydrological Modeling

GIS hydrological models can provide a spatial element that other hydrological models lack, with the analysis of variables such as slope, aspect and watershed or catchment area. Terrain analysis is fundamental to hydrology, since water always flows down a slope. As basic terrain analysis of a digital elevation model (DEM) involves calculation of slope and aspect, DEMs are very useful for hydrological analysis. Slope and aspect can then be used to determine direction of surface runoff, and hence flow accumulation for the formation of streams, rivers and lakes. Areas of divergent flow can also give a clear indication of the boundaries of a catchment. Once a flow direction and accumulation matrix has been created, queries can be performed that show contributing or dispersal areas at a certain point. More detail can be added to the model, such as terrain roughness, vegetation types and soil types, which can influence infiltration and evapotranspiration rates, and hence influencing surface flow. One of the main uses of hydrological modeling is in environmental contamination research. Other applications of hydrological modeling include groundwater and surface water mapping, as well as flood risk maps.

Cartographic Modeling

An example of use of layers in a GIS application. In this example, the forest-cover layer (light green) forms the bottom layer, with the topographic layer (contour lines) over it. Next up is a standing water layer (pond, lake) and then a flowing water layer (stream, river), followed by the boundary layer and finally the road layer on top. The order is very important in order to properly display the final result. Note that the ponds are layered under the streams, so that a stream line can be seen overlying one of the ponds.

Dana Tomlin probably coined the term "cartographic modeling" in his PhD dissertation; he later used it in the title of his book, Geographic Information Systems and Cartographic Modeling. Cartographic modeling refers to a process where several thematic layers of the same area are produced, processed, and analyzed. Tomlin used raster layers, but the overlay method can be used more generally. Operations on map layers can be combined into algorithms, and eventually into simulation or optimization models.

Map Overlay

The combination of several spatial datasets (points, lines, or polygons) creates a new output vector dataset, visually similar to stacking several maps of the same region. These overlays are similar to mathematical Venn diagram overlays. A union overlay combines the geographic features and attribute tables of both inputs into a single new output. An intersect overlay defines the area where both inputs overlap and retains a set of attribute fields for each. A symmetric difference overlay defines an output area that includes the total area of both inputs except for the overlapping area.

Data extraction is a GIS process similar to vector overlay, though it can be used in either vector or raster data analysis. Rather than combining the properties and features of both datasets, data extraction involves using a "clip" or "mask" to extract the features of one data set that fall within the spatial extent of another dataset.

In raster data analysis, the overlay of datasets is accomplished through a process known as "local operation on multiple rasters" or "map algebra", through a function that combines the values of each raster's matrix. This function may weigh some inputs more than others through use of an "index model" that reflects the influence of various factors upon a geographic phenomenon.

Geostatistics

Geostatistics is a branch of statistics that deals with field data, spatial data with a continuous index. It provides methods to model spatial correlation, and predict values at arbitrary locations (interpolation).

Hillshade model derived from a digital elevation model of
the Valestra area in the northern Apennines.

When phenomena are measured, the observation methods dictate the accuracy of any subsequent analysis. Due to the nature of the data (e.g. traffic patterns in an urban environment; weather patterns over the Pacific Ocean), a constant or dynamic degree of precision is always lost in the measurement. This loss of precision is determined from the scale and distribution of the data collection.

To determine the statistical relevance of the analysis, an average is determined so that points (gradients) outside of any immediate measurement can be included to determine their predicted behavior. This is due to the limitations of the applied statistic and data collection methods, and interpolation is required to predict the behavior of particles, points, and locations that are not directly measurable.

Interpolation is the process by which a surface is created, usually a raster dataset, through the input of data collected at a number of sample points. There are several forms of interpolation, each which treats the data differently, depending on the properties of the data set. In comparing interpolation methods, the first consideration should be whether or not the source data will change (exact or approximate). Next is whether the method is subjective, a human interpretation, or objective. Then there is the nature of transitions between points: are they abrupt or gradual. Finally, there is whether a method is global (it uses the entire data set to form the model), or local where an algorithm is repeated for a small section of terrain.

Interpolation is a justified measurement because of a spatial autocorrelation principle that recognizes that data collected at any position will have a great similarity to, or influence of those locations within its immediate vicinity.

Digital elevation models, triangulated irregular networks, edge-finding algorithms, Thiessen polygons, Fourier analysis, (weighted) moving averages, inverse distance weighting, kriging, spline, and trend surface analysis are all mathematical methods to produce interpolative data.

Address Geocoding

Geocoding is interpolating spatial locations (X,Y coordinates) from street addresses or any other spatially referenced data such as ZIP Codes, parcel lots and address locations. A reference theme is required to geocode individual addresses, such as a road centerline file with address ranges. The individual address locations have historically been interpolated, or estimated, by examining address ranges along a road segment. These are usually provided in the form of a table or database. The software will then place a dot approximately where that address belongs along the segment of centerline. For example, an address point of 500 will be at the midpoint of a line segment that starts with address 1 and ends with address 1,000. Geocoding can also be applied against actual parcel data, typically from municipal tax maps. In this case, the result of the geocoding will be an actually positioned space as opposed to an interpolated point. This approach is being increasingly used to provide more precise location information.

Reverse Geocoding

Reverse geocoding is the process of returning an estimated street address number as it relates to a given coordinate. For example, a user can click on a road centerline theme (thus providing a coordinate) and have information returned that reflects the estimated house number. This house number is interpolated from a range assigned to that road segment. If the user clicks at the midpoint of a segment that starts with address 1 and ends with 100, the returned value will be somewhere near 50. Note that reverse geocoding does not return actual addresses, only estimates of what should be there based on the predetermined range.

Multi-criteria Decision Analysis

Coupled with GIS, multi-criteria decision analysis methods support decision-makers in analysing a set of alternative spatial solutions, such as the most likely ecological habitat for restoration, against multiple criteria, such as vegetation cover or roads. MCDA uses decision rules to aggregate the criteria, which allows the alternative solutions to be ranked or prioritised. GIS MCDA may reduce costs and time involved in identifying potential restoration sites.

Data Output and Cartography

Cartography is the design and production of maps, or visual representations of spatial data. The vast majority of modern cartography is done with the help of computers, usually using GIS but production of quality cartography is also achieved by importing layers into a design program to refine it. Most GIS software gives the user substantial control over the appearance of the data.

Cartographic work serves two major functions:

First, it produces graphics on the screen or on paper that convey the results of analysis to the people who make decisions about resources. Wall maps and other graphics can be generated, allowing the viewer to visualize and thereby understand the results of analyses or simulations of potential events. Web Map Servers facilitate distribution of generated maps through web browsers using various implementations of web-based application programming interfaces (AJAX, Java, Flash, etc.).

Second, other database information can be generated for further analysis or use. An example would be a list of all addresses within one mile (1.6 km) of a toxic spill.

Graphic Display Techniques

Traditional maps are abstractions of the real world, a sampling of important elements portrayed on a sheet of paper with symbols to represent physical objects. People who use maps must interpret these symbols. Topographic maps show the shape of land surface with contour lines or with shaded relief.

Today, graphic display techniques such as shading based on altitude in a GIS can make relationships among map elements visible, heightening one's ability to extract and analyze information. For example, two types of data were combined in a GIS to produce a perspective view of a portion of San Mateo County, California.

- The digital elevation model, consisting of surface elevations recorded on a 30-meter horizontal grid, shows high elevations as white and low elevation as black.

- The accompanying Landsat Thematic Mapper image shows a false-color infrared image looking down at the same area in 30-meter pixels, or picture elements, for the same coordinate points, pixel by pixel, as the elevation information.

A GIS was used to register and combine the two images to render the three-dimensional perspective view looking down the San Andreas Fault, using the Thematic Mapper image pixels, but shaded using the elevation of the landforms. The GIS display depends on the viewing point of the observer and time of day of the display, to properly render the shadows created by the sun's rays at that latitude, longitude, and time of day.

An archeochrome is a new way of displaying spatial data. It is a thematic on a 3D map that is applied to a specific building or a part of a building. It is suited to the visual display of heat-loss data.

Spatial ETL

Spatial ETL tools provide the data processing functionality of traditional extract, transform, load (ETL) software, but with a primary focus on the ability to manage spatial data. They provide GIS users with the ability to translate data between different standards and proprietary formats, whilst geometrically transforming the data en route. These tools can come in the form of add-ins to existing wider-purpose software such as spreadsheets.

GIS Data Mining

GIS or spatial data mining is the application of data mining methods to spatial data. Data mining, which is the partially automated search for hidden patterns in large databases, offers great potential benefits for applied GIS-based decision making. Typical applications include environmental monitoring. A characteristic of such applications is that spatial correlation between data measurements require the use of specialized algorithms for more efficient data analysis.

Applications

The implementation of a GIS is often driven by jurisdictional (such as a city), purpose, or application requirements. Generally, a GIS implementation may be custom-designed for an organization. Hence, a GIS deployment developed for an application, jurisdiction, enterprise, or purpose may not be necessarily interoperable or compatible with a GIS that has been developed for some other application, jurisdiction, enterprise, or purpose.

GIS provides, for every kind of location-based organization, a platform to update geographical data without wasting time to visit the field and update a database manually. GIS when integrated with other powerful enterprise solutions like SAP and the Wolfram Language helps creating powerful decision support system at enterprise level.

Many disciplines can benefit from GIS technology. An active GIS market has resulted in lower costs and continual improvements in the hardware and software components of GIS, and usage in the fields of science, government, business, and industry, with applications including real estate, public health, crime mapping, national defense, sustainable development, natural resources, climatology, landscape architecture, archaeology, regional and community planning, transportation and logistics. GIS is also diverging into location-based services, which allows GPS-enabled mobile devices to display their location in relation to fixed objects (nearest restaurant, gas station, fire hydrant) or mobile objects (friends, children, police car), or to relay their position back to a central server for display or other processing.

Open Geospatial Consortium Standards

The Open Geospatial Consortium (OGC) is an international industry consortium of 384 companies, government agencies, universities, and individuals participating in a consensus process to develop publicly available geoprocessing specifications. Open interfaces and protocols defined by OpenGIS Specifications support interoperable solutions that "geo-enable" the Web, wireless and location-based services, and mainstream IT, and empower technology developers to make complex spatial information and services accessible and useful with all kinds of applications. Open Geospatial Consortium protocols include Web Map Service, and Web Feature Service.

GIS products are broken down by the OGC into two categories, based on how completely and accurately the software follows the OGC specifications.

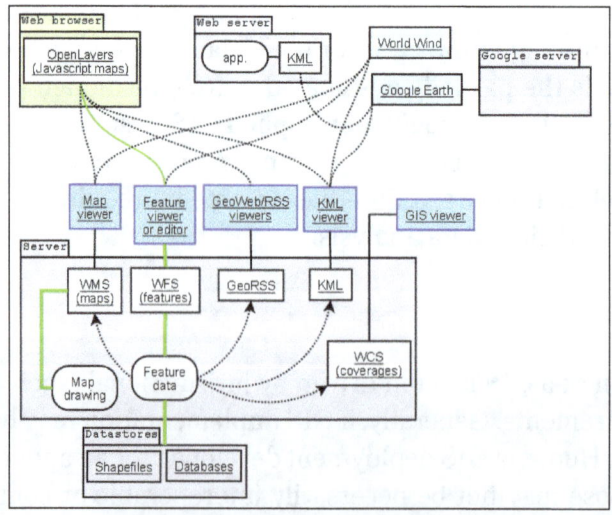

OGC standards help GIS tools communicate.

Compliant Products are software products that comply to OGC's OpenGIS Specifications. When a product has been tested and certified as compliant through the OGC Testing Program, the product is automatically registered as "compliant" on this site.

Implementing Products are software products that implement OpenGIS Specifications but have not yet passed a compliance test. Compliance tests are not available for all specifications. Developers can register their products as implementing draft or approved specifications, though OGC reserves the right to review and verify each entry.

Web Mapping

In recent years there has been a proliferation of free-to-use and easily accessible mapping software such as the proprietary web applications Google Maps and Bing Maps, as well as the free and open-source alternative OpenStreetMap. These services give the public access to huge amounts of geographic data; perceived by many users to be as trustworthy and usable as professional information.

Some of them, like Google Maps and OpenLayers, expose an application programming interface (API) that enable users to create custom applications. These toolkits commonly offer street maps, aerial/satellite imagery, geocoding, searches, and routing functionality. Web mapping has also uncovered the potential of crowdsourcing geodata in projects like OpenStreetMap, which is a collaborative project to create a free editable map of the world. These mashup projects have been proven to provide a high level of value and benefit to end users outside that possible through traditional geographic information.

Adding the Dimension of Time

The condition of the Earth's surface, atmosphere, and subsurface can be examined by feeding satellite data into a GIS. GIS technology gives researchers the ability to examine the variations in Earth processes over days, months, and years. As an example, the changes in vegetation vigor through a growing season can be animated to determine when drought was most extensive in a particular region. The resulting graphic represents a rough measure of plant health. Working with two variables over time would then allow researchers to detect regional differences in the lag between a decline in rainfall and its effect on vegetation.

GIS technology and the availability of digital data on regional and global scales enable such analyses. The satellite sensor output used to generate a vegetation graphic is produced for example by the advanced very-high-resolution radiometer (AVHRR). This sensor system detects the amounts of energy reflected from the Earth's surface across various bands of the spectrum for surface areas of about 1 square kilometer. The satellite sensor produces images of a particular location on the Earth twice a day. AVHRR and more recently the moderate-resolution imaging spectroradiometer (MODIS) are only two of many sensor systems used for Earth surface analysis.

In addition to the integration of time in environmental studies, GIS is also being explored for its ability to track and model the progress of humans throughout their daily routines. A concrete example of progress in this area is the recent release of time-specific population data by the U.S. Census. In this data set, the populations of cities are shown for daytime and evening hours highlighting the pattern of concentration and dispersion generated by North American commuting patterns. The manipulation and generation of data required to produce this data would not have been possible without GIS.

Using models to project the data held by a GIS forward in time have enabled planners to test policy decisions using spatial decision support systems.

Semantics

Tools and technologies emerging from the World Wide Web Consortium's Semantic Web are proving useful for data integration problems in information systems. Correspondingly, such technologies have been proposed as a means to facilitate interoperability and data reuse among GIS applications. and also to enable new analysis mechanisms.

Ontologies are a key component of this semantic approach as they allow a formal, machine-readable specification of the concepts and relationships in a given domain. This in turn allows a GIS to focus on the intended meaning of data rather than its syntax or structure. For example, reasoning that a land cover type classified as deciduous needle-leaf trees in one dataset is a specialization or subset of land cover type forest in another more roughly classified dataset can help a GIS automatically merge the two datasets under the more general land cover classification. Tentative ontologies have been developed in areas related to GIS applications, for example the hydrology ontology developed by the Ordnance Survey in the United Kingdom and the SWEET ontologies developed by NASA's Jet Propulsion Laboratory. Also, simpler ontologies and semantic metadata standards are being proposed by the W3C Geo Incubator Group to represent geospatial data on the web. GeoSPARQL is a standard developed by the Ordnance Survey, United States Geological Survey, Natural Resources Canada, Australia's Commonwealth Scientific and Industrial Research Organisation and others to support ontology creation and reasoning using well-understood OGC literals (GML, WKT), topological relationships (Simple Features, RCC8, DE-9IM), RDF and the SPARQL database query protocols.

Recent research results in this area can be seen in the International Conference on Geospatial Semantics and the Terra Cognita – Directions to the Geospatial Semantic Web workshop at the International Semantic Web Conference.

Implications of GIS in Society

With the popularization of GIS in decision making, scholars have begun to scrutinize the social and political implications of GIS. GIS can also be misused to distort reality for individual and political gain. It has been argued that the production, distribution, utilization, and representation of geographic information are largely related with the social context and has the potential to increase citizen trust in government. Other related topics include discussion on copyright, privacy, and censorship. A more optimistic social approach to GIS adoption is to use it as a tool for public participation.

GIS in Education

At the end of the 20th century, GIS began to be recognized as tools that could be used in the classroom. The benefits of GIS in education seem focused on developing spatial thinking, but there is not enough bibliography or statistical data to show the concrete scope of the use of GIS in education around the world, although the expansion has been faster in those countries where the curriculum mentions them.

GIS seem to provide many advantages in teaching geography because they allow for analyses based on real geographic data and also help raise many research questions from teachers and students in classrooms, as well as they contribute to improvement in learning by developing spatial and geographical thinking and, in many cases, student motivation.

GIS in Local Government

GIS is proven as an organization-wide, enterprise and enduring technology that continues to change how local government operates. Government agencies have adopted GIS technology as a method to better manage the following areas of government organization:

- Economic Development departments use interactive GIS mapping tools, aggregated with other data (demographics, labor force, business, industry, talent) along with a database of available commercial sites and buildings in order to attract investment and support existing business. Businesses making location decisions can use the tools to choose communities and sites that best match their criteria for success. GIS Planning's ZoomProspector Enterprise an Intelligence Components software is the industry leader, servicing more than 60% of the US population, more than 30% of Canadians, and locations in the UK and Switzerland Public. Safety operations such as Emergency Operations Centers, Fire Prevention, Police and Sheriff mobile technology and dispatch, and mapping weather risks.

- Parks and Recreation departments and their functions in asset inventory, land conservation, land management, and cemetery management.

- Public Works and Utilities, tracking water and stormwater drainage, electrical assets, engineering projects, and public transportation assets and trends.

- Fiber Network Management for interdepartmental network assets.

- School analytical and demographic data, asset management, and improvement/expansion planning.

- Public Administration for election data, property records, and zoning/management.

The Open Data initiative is pushing local government to take advantage of technology such as GIS technology, as it encompasses the requirements to fit the Open Data/Open Government model of transparency. With Open Data, local government organizations can implement Citizen Engagement applications and online portals, allowing citizens to see land information, report potholes and signage issues, view and sort parks by assets, view real-time crime rates and utility repairs, and much more. The push for open data within government organizations is driving the growth in local government GIS technology spending, and database management.

REMOTE SENSING

Remote sensing is the acquisition of information about an object or phenomenon without making physical contact with the object and thus in contrast to on-site observation, especially the Earth. Remote sensing is used in numerous fields, including geography, land surveying and most Earth science disciplines (for example, hydrology, ecology, meteorology, oceanography, glaciology, geology); it also has military, intelligence, commercial, economic, planning, and humanitarian applications.

In current usage, the term "remote sensing" generally refers to the use of satellite- or aircraft-based sensor technologies to detect and classify objects on Earth, including on the surface and in the atmosphere and oceans, based on propagated signals (e.g. electromagnetic radiation). It may be split into "active" remote sensing (such as when a signal is emitted by a satellite or aircraft and its reflection by the object is detected by the sensor) and "passive" remote sensing (such as when the reflection of sunlight is detected by the sensor).

This video is about how Landsat was used to identify areas of conservation in the Democratic Republic of the Congo, and how it was used to help map an area called MLW in the north.

Passive sensors gather radiation that is emitted or reflected by the object or surrounding areas. Reflected sunlight is the most common source of radiation measured by passive sensors. Examples of passive remote sensors include film photography, infrared, charge-coupled devices, and radiometers. Active collection, on the other hand, emits energy in order to scan objects and areas whereupon a sensor then detects and measures the radiation that is reflected or backscattered from the target. RADAR and LiDAR are examples of active remote sensing where the time delay between emission and return is measured, establishing the location, speed and direction of an object.

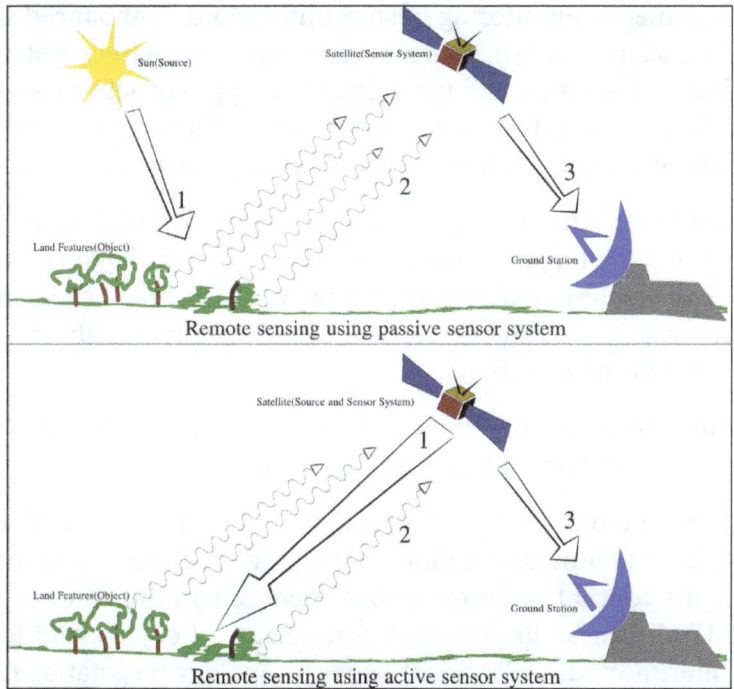

Illustration of remote sensing.

Remote sensing makes it possible to collect data of dangerous or inaccessible areas. Remote sensing applications include monitoring deforestation in areas such as the Amazon Basin, glacial features in Arctic and Antarctic regions, and depth sounding of coastal and ocean depths. Military collection during the Cold War made use of stand-off collection of data about dangerous border areas. Remote sensing also replaces costly and slow data collection on the ground, ensuring in the process that areas or objects are not disturbed.

Orbital platforms collect and transmit data from different parts of the electromagnetic spectrum, which in conjunction with larger scale aerial or ground-based sensing and analysis, provides researchers with enough information to monitor trends such as El Niño and other natural long and short term phenomena. Other uses include different areas of the earth sciences such as natural resource management, agricultural fields such as land usage and conservation, and national security and overhead, ground-based and stand-off collection on border areas.

Types of Data Acquisition Techniques

The basis for multispectral collection and analysis is that of examined areas or objects that reflect or emit radiation that stand out from surrounding areas.

Applications of Remote Sensing

- Conventional radar is mostly associated with aerial traffic control, early warning, and certain large scale meteorological data. Doppler radar is used by local law enforcements' monitoring of speed limits and in enhanced meteorological collection such as wind speed and direction within weather systems in addition to precipitation location and intensity. Other types of active collection includes plasmas in the ionosphere. Interferometric synthetic aperture radar is used to produce precise digital elevation models of large scale terrain.

- Laser and radar altimeters on satellites have provided a wide range of data. By measuring the bulges of water caused by gravity, they map features on the seafloor to a resolution of a mile or so. By measuring the height and wavelength of ocean waves, the altimeters measure wind speeds and direction, and surface ocean currents and directions.

- Ultrasound (acoustic) and radar tide gauges measure sea level, tides and wave direction in coastal and offshore tide gauges.

- Light detection and ranging (LIDAR) is well known in examples of weapon ranging, laser illuminated homing of projectiles. LIDAR is used to detect and measure the concentration of various chemicals in the atmosphere, while air-borne LIDAR can be used to measure heights of objects and features on the ground more accurately than with radar technology. Vegetation remote sensing is a principal application of LIDAR.

- Radiometers and photometers are the most common instrument in use, collecting reflected and emitted radiation in a wide range of frequencies. The most common are visible and infrared sensors, followed by microwave, gamma ray and rarely, ultraviolet. They may also be used to detect the emission spectra of various chemicals, providing data on chemical concentrations in the atmosphere.

- Radiometers are also used at night, because artificial light emissions are a key signature of human activity. Applications include remote sensing of population, GDP, and damage to infrastructure from war or disasters.

- Spectropolarimetric Imaging has been reported to be useful for target tracking purposes by researchers at the U.S. Army Research Laboratory. They determined that manmade items possess polarimetric signatures that are not found in natural objects. These conclusions were drawn from the imaging of military trucks, like the Humvee, and trailers with their acousto-optic tunable filter dual hyperspectral and spectropolarimetric VNIR Spectropolarimetric Imager.

- Stereographic pairs of aerial photographs have often been used to make topographic maps by imagery and terrain analysts in trafficability and highway departments for potential routes, in addition to modelling terrestrial habitat features.

- Simultaneous multi-spectral platforms such as Landsat have been in use since the 1970s. These thematic mappers take images in multiple wavelengths of electro-magnetic radiation (multi-spectral) and are usually found on Earth observation satellites, including (for example) the Landsat program or the IKONOS satellite. Maps of land cover and land use from thematic mapping can be used to prospect for minerals, detect or monitor land usage, detect invasive vegetation, deforestation, and examine the health of indigenous plants and crops, including entire farming regions or forests. Prominent scientists using remote sensing for this purpose include Janet Franklin and Ruth DeFries. Landsat images are used by regulatory agencies such as KYDOW to indicate water quality parameters including Secchi depth, chlorophyll a density and total phosphorus content. Weather satellites are used in meteorology and climatology.

- Hyperspectral imaging produces an image where each pixel has full spectral information with imaging narrow spectral bands over a contiguous spectral range. Hyperspectral imagers are used in various applications including mineralogy, biology, defence, and environmental measurements.

- Within the scope of the combat against desertification, remote sensing allows researchers to follow up and monitor risk areas in the long term, to determine desertification factors, to support decision-makers in defining relevant measures of environmental management, and to assess their impacts.

Geodetic

- Geodetic remote sensing can be gravimetric or geometric. Overhead gravity data collection was first used in aerial submarine detection. This data revealed minute perturbations in the Earth's gravitational field that may be used to determine changes in the mass distribution of the Earth, which in turn may be used for geophysical studies, as in GRACE. Geometric remote sensing includes position and deformation imaging using InSAR, LIDAR, etc.

Acoustic and Near-acoustic

- Sonar: Passive sonar, listening for the sound made by another object (a vessel, a whale etc.); active sonar, emitting pulses of sounds and listening for echoes, used for detecting, ranging and measurements of underwater objects and terrain.

- Seismograms taken at different locations can locate and measure earthquakes (after they occur) by comparing the relative intensity and precise timings.

- Ultrasound: Ultrasound sensors, that emit high frequency pulses and listening for echoes, used for detecting water waves and water level, as in tide gauges or for towing tanks.

To coordinate a series of large-scale observations, most sensing systems depend on the following: platform location and the orientation of the sensor. High-end instruments now often use positional information from satellite navigation systems. The rotation and orientation is often provided within a degree or two with electronic compasses.

Compasses can measure not just azimuth (i. e. degrees to magnetic north), but also altitude (degrees above the horizon), since the magnetic field curves into the Earth at different angles at different latitudes. More exact orientations require gyroscopic-aided orientation, periodically realigned by different methods including navigation from stars or known benchmarks.

Data Characteristics

The quality of remote sensing data consists of its spatial, spectral, radiometric and temporal resolutions.

- Spatial resolution: The size of a pixel that is recorded in a raster image – typically pixels may correspond to square areas ranging in side length from 1 to 1,000 metres (3.3 to 3,280.8 ft).

- Spectral resolution: The wavelength of the different frequency bands recorded – usually, this is related to the number of frequency bands recorded by the platform. Current Landsat collection is that of seven bands, including several in the infrared spectrum, ranging from a spectral resolution of 0.7 to 2.1 μm. The Hyperion sensor on Earth Observing-1 resolves 220 bands from 0.4 to 2.5 μm, with a spectral resolution of 0.10 to 0.11 μm per band.

- Radiometric resolution: The number of different intensities of radiation the sensor is able to distinguish. Typically, this ranges from 8 to 14 bits, corresponding to 256 levels of the gray scale and up to 16,384 intensities or "shades" of colour, in each band. It also depends on the instrument noise.

- Temporal resolution: The frequency of flyovers by the satellite or plane, and is only relevant in time-series studies or those requiring an averaged or mosaic image as in deforesting monitoring. This was first used by the intelligence community where repeated coverage revealed changes in infrastructure, the deployment of units or the modification/introduction of equipment. Cloud cover over a given area or object makes it necessary to repeat the collection of said location.

Data Processing

In order to create sensor-based maps, most remote sensing systems expect to extrapolate sensor data in relation to a reference point including distances between known points on the ground. This depends on the type of sensor used. For example, in conventional photographs, distances are accurate in the center of the image, with the distortion of measurements increasing the farther you get from the center. Another factor is that of the platen against which the film is pressed can cause severe errors when photographs are used to measure ground distances. The step in which this problem is resolved is called georeferencing, and involves computer-aided matching of points in the image (typically 30 or more points per image) which is extrapolated with the use of an established benchmark, "warping" the image to produce accurate spatial data. As of the early 1990s, most satellite images are sold fully georeferenced.

In addition, images may need to be radiometrically and atmospherically corrected.

- Radiometric correction: Allows avoidance of radiometric errors and distortions. The illumination of objects on the Earth surface is uneven because of different properties of the relief. This factor is taken into account in the method of radiometric distortion correction. Radiometric correction gives a scale to the pixel values, e. g. the monochromatic scale of 0 to 255 will be converted to actual radiance values.

- Topographic correction (also called terrain correction): In rugged mountains, as a result of terrain, the effective illumination of pixels varies considerably. In a remote sensing image, the pixel on the shady slope receives weak illumination and has a low radiance value, in contrast, the pixel on the sunny slope receives strong illumination and has a high radiance value. For the same object, the pixel radiance value on the shady slope will be different from that on the sunny slope. Additionally, different objects may have similar radiance values. These ambiguities seriously affected remote sensing image information extraction accuracy in mountainous areas. It became the main obstacle to further application of remote sensing images. The purpose of topographic correction is to eliminate this effect, recovering the true reflectivity or radiance of objects in horizontal conditions. It is the premise of quantitative remote sensing application.

- Atmospheric correction: Elimination of atmospheric haze by rescaling each frequency band so that its minimum value (usually realised in water bodies) corresponds to a pixel value of 0. The digitizing of data also makes it possible to manipulate the data by changing gray-scale values.

Interpretation is the critical process of making sense of the data. The first application was that of aerial photographic collection which used the following process; spatial measurement through the use of a light table in both conventional single or stereographic coverage, added skills such as the use of photogrammetry, the use of photomosaics, repeat coverage, Making use of objects' known dimensions in order to detect modifications. Image Analysis is the recently developed automated computer-aided application which is in increasing use.

Object-Based Image Analysis (OBIA) is a sub-discipline of GIScience devoted to partitioning remote sensing (RS) imagery into meaningful image-objects, and assessing their characteristics through spatial, spectral and temporal scale.

Old data from remote sensing is often valuable because it may provide the only long-term data for a large extent of geography. At the same time, the data is often complex to interpret, and bulky to store. Modern systems tend to store the data digitally, often with lossless compression. The difficulty with this approach is that the data is fragile, the format may be archaic, and the data may be easy to falsify. One of the best systems for archiving data series is as computer-generated machine-readable ultrafiche, usually in typefonts such as OCR-B, or as digitized half-tone images. Ultrafiches survive well in standard libraries, with lifetimes of several centuries. They can be created, copied, filed and retrieved by automated systems. They are about as compact as archival magnetic media, and yet can be read by human beings with minimal, standardized equipment.

Generally speaking, remote sensing works on the principle of the inverse problem: while the object or phenomenon of interest (the state) may not be directly measured, there exists some other variable that can be detected and measured (the observation) which may be related to the object of interest through a calculation. The common analogy given to describe this is trying to determine the type of animal from its footprints. For example, while it is impossible to directly measure temperatures in the upper atmosphere, it is possible to measure the spectral emissions from a known chemical species (such as carbon dioxide) in that region. The frequency of the emissions may then be related via thermodynamics to the temperature in that region.

Data Processing Levels

To facilitate the discussion of data processing in practice, several processing "levels" were first defined in 1986 by NASA as part of its Earth Observing System and steadily adopted since then, both internally at NASA (e. g.,) and elsewhere (e. g.,); these definitions are:

Level	Description
0	Reconstructed, unprocessed instrument and payload data at full resolution, with any and all communications artifacts (e. g., synchronization frames, communications headers, duplicate data) removed.
1a	Reconstructed, unprocessed instrument data at full resolution, time-referenced, and annotated with ancillary information, including radiometric and geometric calibration coefficients and georeferencing parameters (e. g., platform ephemeris) computed and appended but not applied to the Level 0 data (or if applied, in a manner that level 0 is fully recoverable from level 1a data).
1b	Level 1a data that have been processed to sensor units (e. g., radar backscatter cross section, brightness temperature, etc.); not all instruments have Level 1b data; level 0 data is not recoverable from level 1b data.
2	Derived geophysical variables (e. g., ocean wave height, soil moisture, ice concentration) at the same resolution and location as Level 1 source data.
3	Variables mapped on uniform spacetime grid scales, usually with some completeness and consistency (e. g., missing points interpolated, complete regions mosaicked together from multiple orbits, etc.).
4	Model output or results from analyses of lower level data (i. e., variables that were not measured by the instruments but instead are derived from these measurements).

A Level 1 data record is the most fundamental (i. e., highest reversible level) data record that has significant scientific utility, and is the foundation upon which all subsequent data sets are produced. Level 2 is the first level that is directly usable for most scientific applications; its value is much greater than the lower levels. Level 2 data sets tend to be less voluminous than Level 1 data because they have been reduced temporally, spatially, or spectrally. Level 3 data sets are generally smaller than lower level data sets and thus can be dealt with without incurring a great deal of data handling overhead. These data tend to be generally more useful for many applications. The regular spatial and temporal organization of Level 3 datasets makes it feasible to readily combine data from different sources.

While these processing levels are particularly suitable for typical satellite data processing pipelines, other data level vocabularies have been defined and may be appropriate for more heterogeneous workflows.

Software

Remote sensing data are processed and analyzed with computer software, known as a remote sensing application. A large number of proprietary and open source applications exist to process remote sensing data. Remote sensing software packages include:

- ERDAS IMAGINE from Hexagon Geospatial (Separated from Intergraph SG&I),
- ENVI from Harris GeospatialSolutions,
- PCI Geomatica,
- TNTmips from MicroImages,
- IDRISI from Clark Labs,
- eCognition from Trimble,
- RemoteView made by Overwatch Textron Systems,
- Dragon/ips is one of the oldest remote sensing packages still available, and is in some cases free.

Open source remote sensing software includes:

- Opticks (software),
- Orfeo toolbox,
- Sentinel Application Platform (SNAP) from the European Space Agency (ESA),
- Others mixing remote sensing and GIS capabilities are: GRASS GIS, ILWIS, QGIS, and TerraLook.

According to Research, the most used applications among Asian academic groups involved in remote sensing are as follows: ERDAS 36% (ERDAS IMAGINE 25% & ER-Mapper 11%); ESRI 30%; ITT Visual Information Solutions ENVI 17%; MapInfo 17%.

Among Western Academic respondents as follows: ESRI 39%, ERDAS IMAGINE 27%, MapInfo 9%, and AutoDesk 7%.

In education, those that want to go beyond simply looking at satellite images print-outs either use general remote sensing software (e.g. QGIS), Google Earth, StoryMaps or a software/web-app developed specifically for education.

GEOSTATISTICS

Geostatistics is a branch of statistics focusing on spatial or spatiotemporal datasets. Developed originally to predict probability distributions of ore grades for mining operations, it is currently applied in diverse disciplines including petroleum geology, hydrogeology, hydrology, meteorology, oceanography, geochemistry, geometallurgy, geography, forestry, environmental control, landscape ecology, soil science, and agriculture (esp. in precision farming). Geostatistics is applied in varied branches of geography, particularly those involving the spread of diseases (epidemiology), the practice of commerce and military planning (logistics), and the development of efficient spatial networks. Geostatistical algorithms are incorporated in many places, including geographic information systems (GIS) and the R statistical environment.

Geostatistics is intimately related to interpolation methods, but extends far beyond simple interpolation problems. Geostatistical techniques rely on statistical models that are based on random function (or random variable) theory to model the uncertainty associated with spatial estimation and simulation.

A number of simpler interpolation methods/algorithms, such as inverse distance weighting, bilinear interpolation and nearest-neighbor interpolation, were already well known before geostatistics. Geostatistics goes beyond the interpolation problem by considering the studied phenomenon at unknown locations as a set of correlated random variables.

Let $Z(x)$ be the value of the variable of interest at a certain location x. This value is unknown (e.g. temperature, rainfall, piezometric level, geological facies, etc.). Although there exists a value at location x that could be measured, geostatistics considers this value as random since it was not measured, or has not been measured yet. However, the randomness of $Z(x)$ is not complete, but defined by a cumulative distribution function (CDF) that depends on certain information that is known about the value $Z(x)$:

$$F(z, x) = \text{Prob}\{Z(x) \leqslant z \mid \text{information}\}$$

Typically, if the value of Z is known at locations close to x (or in the neighborhood of x) one can constrain the CDF of $Z(x)$ by this neighborhood: if a high spatial continuity is assumed, $Z(x)$ can only have values similar to the ones found in the neighborhood.

Conversely, in the absence of spatial continuity Z(x) can take any value. The spatial continuity of the random variables is described by a model of spatial continuity that can be either a parametric function in the case of variogram-based geostatistics, or have a non-parametric form when using other methods such as multiple-point simulation or pseudo-genetic techniques.

By applying a single spatial model on an entire domain, one makes the assumption that Z is a stationary process. It means that the same statistical properties are applicable on the entire domain. Several geostatistical methods provide ways of relaxing this stationarity assumption.

In this framework, one can distinguish two modeling goals:

- Estimating the value for Z(x), typically by the expectation, the median or the mode of the CDF f(z,x). This is usually denoted as an estimation problem.

- Sampling from the entire probability density function f(z,x) by actually considering each possible outcome of it at each location. This is generally done by creating several alternative maps of Z, called realizations. Consider a domain discretized in N grid nodes (or pixels). Each realization is a sample of the complete N-dimensional joint distribution function:

$$F(z,x) = \text{Prob}\{Z(x_1) \leqslant z_1, Z(x_2) \leqslant z_2, ..., Z(x_N) \leqslant z_N\}$$

In this approach, the presence of multiple solutions to the interpolation problem is acknowledged. Each realization is considered as a possible scenario of what the real variable could be. All associated workflows are then considering ensemble of realizations, and consequently ensemble of predictions that allow for probabilistic forecasting. Therefore, geostatistics is often used to generate or update spatial models when solving inverse problems.

References

- Cartography, science: britannica.com, Retrieved 18 March, 2020

- Schowengerdt, Robert A. (2007). Remote sensing: models and methods for image processing (3rd ed.). Academic Press. p. 2. ISBN 978-0-12-369407-2

- What-is-geoinformatics, general-knowledge-1556110951-1: jagranjosh.com, Retrieved 12 January, 2020

- Maliene V, Grigonis V, Palevičius V, Griffiths S (2011). "Geographic information system: Old principles with new capabilities". Urban Design International. 16 (1): 1–6. doi:10.1057/udi.2010.25

- Aerial-photography-meaning-and-interpretation-geography, topography-aerial-photography-5964: geographynotes.com, Retrieved 15 February, 2020

- Bolstad, Paul. GIS Fundamentals (PDF) (5th ed.). Atlas books. p. 102. ISBN 978-0-9717647-3-6

- McGovern, Eugene A.; Holden, Nicholas M.; Ward, Shane M.; Collins, James F. (January 2000). "Remotely sensed satellite imagery as an information source for industrial peatlands management". Resources, Conservation and Recycling. 28 (1–2): 67–83. doi:10.1016/s0921-3449(99)00034-8. ISSN 0921-3449

5

Diverse Aspects of Geography

Some of the diverse aspects that fall under the field of geography are geoarchaeology, Tobler's first law of geography, glacial refugium, geographic contiguity, geo-literacy, geography of aging, etc. The topics elaborated in this chapter will help in gaining a better perspective of all the related aspects of geography.

TOBLER'S FIRST LAW OF GEOGRAPHY

Tobler's first law of geography was originally "promulgated in a paper published in the journal Economic Geography. Tobler's article described a model of population growth for Detroit from 1910 to 2000. The model produced a simulation that was presented as a wire-frame, three-dimensional block diagram. Tobler suggested that if the simulated surfaces were viewed at 16 frames per second then a new surface would be generated for every month of the simulated time-period, maximizing viewing effectiveness. Many of the spatial characteristics of the model were provided in a paper published a year earlier and cited in the 1970 article. A total of 1150 grid cells were used, each being 1.5 miles square. To operationalize his simulation, Tobler stated that he would "invoke the first law of geography: everything is related to everything else, but near things are more related than distant things". Tobler had earlier argued that "everything is related to everything else" but he added the additional proviso to make the model more localized and thus more tractable. Temporal influences on the population at each stage of the simulation came from the population during the previous decades, declining by 50% for each passing decade. Spatial influence came from the adjacent cells and this influence was modeled.

Tobler's first law have been immensely influential. His 1970 paper has now been cited over 5000 times in the academic and research literature. In June 2003, a Google search of "Tobler's first law of geography" produced "at least 150 returns", but in April 2016, the same search produced almost 8000 returns and in August 2017 there were 19000 returns. Interest in TFL appears to have been growing exponentially.

The Law Unmasked, Deconstructed and Parsed

To understand the importance of Tobler's law we must question what exactly a first law in the discipline of geography might imply, and ask if other academic disciplines have laws and, more specifically, if they are the most important laws, that is, "first laws". We should ask whether this was the "first" such law in geography. We need to know what is meant by "everything". We need to ask how we determine whether two "things" are "related" and what we mean by a "thing". Finally, we should make explicit how we measure distance and what is "near" as opposed to being more "distant" and if the re-latedness ever drops to zero.

Determining the Importance and Law-like Status of TFL

When Tobler said that this was "the first law of geography," he was not suggesting that this was the first law to be declared. He was arguing that this was the most important law in geography and that distance was the most important variable governing the in-fluence of one entity on another. In his case study of Detroit, the variable of interest (the "thing") was persons aggregated into populations, but the implication of TFL is that most geographic variables would show a distance-decay effect.

It is important to determine whether others had suggested there should be laws in geography. Therefore, the question must be asked whether this was the first time any scholar had suggested that there be a "first law" in geography. The answer is that possibly it was, but in the years before 1970 there had been an intense interest among scholars in defining laws and theorems in geography. This concern for formulating laws was a consequence of the increasingly bitter debate between Richard Hartshorne and William Bunge about the nature of geography. Barnes provides details of their differing views. Bunge had studied at the University of Wisconsin for his MS degree with Hartshorne but when Bunge moved on to his doctoral program it was Hartshorne who cast a negative vote, failing Bunge in his PhD Prelims exam. Hartshorne maintained that geography's role was to describe the uniqueness of geographical regions and that, according to Lewis, "geographers study individual cases rather than construct scientific laws". Bunge argued that entities studied in geography were not unique, in the sense that they had similarities that could be studied and characterized and that these similarities could be measured and exploited allowing for the construction of functional models that produced informative explanations and predictions, and therefore a scientific approach to geography.

Bunge maintained that the traditional science disciplines did not always seek to argue that cause and effect or deterministic laws were essential. Thus, it might be reasoned that a stochastic law could yield useful predictions in a more scientific, mathematically formulated approach to geography. Ironically, distance played a vital role in the debate. In the years between the founding of the first independent Department of Geography at the University of Chicago in 1903 and the 1950s, when other departments of

Geography had been established across the Mid-West Hartshorne's influence dominated the heartland of geography. Hartshorne's view emphasized regional, area studies and a descriptive, idiographic approach dominated the discipline. Weiner has suggested that the ancient Greek practice of ostracism (banishment) from the Athens agora, their meeting place of intellectual activity, might have stimulated those so punished, such as Thucydides, to even greater creative achievements. This may have been true when Bunge moved to the margins of Geography, to a location where Hartshorne might have been supposed to have no influence. This was the Department of Geography at the University of Washington in Seattle. Besides being one of the most geographically remote locations from Madison in the contiguous US, the department was pioneering a scientific, mathematical, nomothetic approach to Geography. Spatial, statistical, mathematical laws were now the Holy Grail. William Garrison was the faculty mentor of a group of graduate students, who were dubbed "Garrison's Raiders" and who espoused this new paradigm, a coterie that included Bunge and Waldo Tobler and represented a new agora for both of these iconoclasts.

At the AAG Forum, Miller endorsed this approach to science and the creation of law-like statements arguing "that science accepts the concept of empirical laws, or compact descriptions of patterns and regularities ... [that] are not required to be immutable truths". Phillips also argued for a less rigid definition of a "law," citing Cole and King's classic text, Quantitative Geography, which provided six definitions of a law that were "used in geography, several of which would apply to TFL". Neither Miller nor Phillips was particularly concerned with the fact that Tobler claimed his proposition to be a "law," but for Smith this was the primary concern. He argued that had Tobler claimed his proposition to be a rule rather than a law then it would have attracted little attention and certainly would not have generated the need for an AAG Forum to debate it. Smith did not accept that empirical generalizations could or should be regarded as laws. He contended that there are three requirements for a law: "it must be universal, synthetic and necessary". Citing David Harvey's well-known text, Explanation in Geography, Smith stated that universality requires that a law must always apply to all members of a class. It must join two concepts, a subject and a predicate which states something about the subject (the synthetic requirement). The necessary requirement was that it is not simply an "accidental" relationship, which in the case of TFL would imply that distance always represents decreasing utility. Smith claimed that although increasing distance often represents a disutility sometimes this is just not true. Examples abound of increasing distance being a benefit (e.g., the "Not In My Back Yard," or NIMBY, effect on location of noxious facilities) or the lack of influence in social networks, and so the concept of what is a law, in Smith's view, was debased. Tobler addressed the NIMBY issue directly noting that spatial autocorrelation does not always have to be positive.

Tobler's paper does not necessarily imply a contradiction between the world views of Hartshorne and Bunge. Tobler stated that he was interested only in birth, death, and migration processes and not in the reasons why an individual is born in a given time period or why that individual dies or migrates. Such influences are likely to vary with the uniqueness of each individual. Using the "first law," Tobler created a simple model that has high generality (applicability to cities other than Detroit), high explanatory power (it replicated the historical population growth processes in the city of Detroit), and that it simulated future population change in Detroit (verifiable, but only with each passing census).

Measuring Distance

To understand TFL it is important to recognize that there are many ways to measure distance. Geographers long have recognized that models and analyses can be expressed in both Euclidean and network space. From their earliest days, location-allocation models have been formulated in both types of space. For example, Leon Cooper formulated his solutions in Euclidean or continuous space while Seifollah Louis Hakimi determined optimal facility locations on a graph or network in discrete space. Most of the classic, theoretical models of human geography were formulated in continuous space. More recently, due to the availability of software such as SANET (Spatial Analysis on a Network), Mathematica and the network analyst routines in ArcGIS, human geographers have evaluated distances along networks such as transportation routes. SANET permits network autocorrelations to be evaluated allowing the researcher to determine if "near things," such as traffic accident densities, are "related" or spatially autocorrelated along the links of a transportation system.

Regardless of whether we choose Euclidean distance or network distance, it is crucial to be able to discriminate between what is a near thing and what is a more distant thing. Measuring distance is much more complicated than simply making a choice between continuous space and discrete space. It is so much more complicated that in 2006 Michel Marie and Elena Deza produced a comprehensive and exhaustive Dictionary of Distance. This was quickly replaced, in 2009, by an even more extensive survey, the Encyclopedia of Distances (EOD). Deza and Deza demonstrate that measuring or estimating the distance between two points is of fundamental concern to researchers in a wide variety of disciplines and that these disciplines include those that are interested in using estimates of distance in the physical world where distance may be measured in spatial units (e.g., miles) or in time or cost of travel, and those disciplines that use distance as a measure of similarity or relatedness in a virtual or variable space (the latter being commonly used for finding clusters of observations). It is thus the inverse of Tobler's Law which might be stated: "All things can be located on a map so that similar things are placed closer together than dissimilar things". The "map" may be 1, 2, 3 or N dimensional. Maintaining the similarities (or dissimilarities i.e. distances) from each thing or object to everyone other thing or object that you are trying to spatialize/map becomes easier as the dimensionality of the space increases. Tobler himself tried to re-spatialize a set of pre-Hittite settlements for which he had trade data. His hypothesis was that the more the settlements traded with each other the more similar they were in terms of their location (most of their locations were unknown). He used the gravity model to calculate the distances (similarity) between each of the settlements and then used multidimensional scaling to recreate a map of their locations as these Assyrian merchant colonies might have existed between 1940 BCE and 1740 BCE. In this sense, he was inverting his own law.

Measuring Similarity, Dissimilarity and Association in Space

When determining that "near things are more related than distant things" a researcher must be able to establish this relationship using an objective, reproducible statistic that measures "relatedness". This statistic might be some form of correlation coefficient, such as Pearson's product moment correlation coefficient, r, or Spearman's rank correlation coefficient, r_s ; a measure of association, such as Goodman and Kruskal's λ or Cramer's V; or a measure of dissimilarity between two locations, x and y, determined as the sum of the values obtained from their differences on each variable i standardized to have a mean of zero and a standard deviation of 1.0, to insure that each variable has potentially the same influence in calculating the dissimilarity (relatedness) between the locations. With correlations and measures of association (i.e., measures of similarity), a larger number indicates that the two locations are more positively related. With dissimilarity measures, a larger number means that the two locations are less "related," that is, further apart in a single or multivariable space. Correlations and measures of association may be used as measures of strength of the relatedness of a variable or variables at two locations, and are usually constrained to a range of +1.0 through 0.0 to −1.0 or from +1.0 through to 0.0, allowing for easy comparisons of, in this case, the relatedness between different pairs of locations. Correlations and measures of association are commonly accompanied by an associated statistical test (for r and r_s this is a t test and for λ and V a chi-squared test or G test) allowing the researcher to determine whether an observed statistic is statistically significant or not.

Gould was the first to cite TFL in a paper, in the same issue of Economic Geography. He noted a problem with traditional measures of relatedness, when used with spatial data, is that they frequently violated the assumption of independence. Indeed, TFL states that spatial data are defined by their spatial dependence or spatial autocorrelation. Researchers in geography have had two responses to this conundrum. We can use TFL and the spatial dependence in our data as an opportunity to model spatial autocorrelation as in kriging or by calculating and interpreting local indicators of spatial association (LISA) statistics to gain insights as to what this implies for our understanding of the variables under investigation. Alternatively, we can remove the spatial autocorrelation using more recent methodologies such as spatial regressions, as exemplified in software packages such as GeoDa. An alternative approach to the modeling of near things is to use geographically weighted regression (GWR) models (and, more recently, geographically weighted discriminant analysis (GWDA) and geographically weighted principal component analysis (GWPCA)), which allow the researcher to model local influences using distance or a predefined number of points that specify exactly what is "near" and what is "related".

The Center for Spatial Studies at the University of California at Santa Barbara maintains a webpage on spatial statistics and spatial econometrics software resources including GeoDa. For R-coded software packages that perform spatial statistical analysis, researchers can access the Comprehensive R Archive Network (CRAN) where they can download the R software for various operating systems (Linux, OS X, and Windows) as well as the CRAN Task View for the analysis of spatial data, a website maintained by Roger Bivand. This is the most comprehensive set of routines for the analysis of spatial relatedness available anywhere and includes R code for most of the procedures mentioned here. Spatial autocorrelation is thus both a problem and an opportunity for the researcher using geographical data, but with so many new spatial analytical software tools now available and often embedded in geographic information system (GIS) packages such as ArcGIS, the former has become increasingly less of a concern.

Tobler's First Law and the Classic Models of Human and Urban Geography

The classic models of human geography including Von Thünen's model of agricultural activity around a centralized market town in an isolated state, Christaller's central place theory (CPT) defining the settlement geography of hierarchical urban systems, and Lösch's The Economics of Location, all produce economic landscapes where TFL is consistently shown to be insufficient. Distance is important in Von Thünen's model, but the model produces rings of agricultural specialization that are homogeneous internally but where the land use between adjacent rings (land-use patches which are indeed very "near things") may be quite unrelated. CPT produces a hierarchy of settlements superimposed on a background of uniform agricultural activity. Therefore a large urban center will be located within a rural setting and the largest centers will be spaced the furthest apart, and it is these centers where similar "high-order" goods and services will be found. A hierarchy is inimical to TFL, at least in terms of geographical distance, and so are borders and boundaries. Lösch's analysis produced sectors of economic activity which alternated between high and low densities of settlements and thus, according to Lösch's theoretical analysis, adjacent sectors would be less similar in this respect and, presumably, in their population and other socioeconomic characteristics.

Geographers and other social scientists have identified several models to characterize the internal structure of cities. The best known of these are the concentric ring model of Burgess, the sector model of Hoyt, and the multiple nuclei model of Harris and Ullman, all three of which were reconciled by Bourne and Murdie's factor analysis of the internal structure of the city of Toronto, Canada. It might be argued that TFL applies within the Burgess rings, or within Hoyt's sectors, or inside Harris and Ullman's ethnically homogeneous nuclei, but again, on the borders of these zones, there might be radically different residential communities or zones of economic activity living adjacent to each other. These inconsistencies in the validity of TFL are in part a function of scale. When a researcher concentrates on the "big picture" and "zooms out" from the map, variables that exhibited a gradual decline with increasing distance from a given location may, at a more detailed scale, simply appear as uniform concentrations with sharp boundaries where the spatial autocorrelation declines to zero. Thus first-order trends may appear as second-order concentrations.

Less theoretical, more applied, empirical models have fared better, especially those used in transportation geography. Step 2 of the four-step transportation planning model is based on the premise that commuters moving out of trip production zones, where they live, and toward trip attraction zones where they work will attempt to minimize the length of their journey to work. Trip distributions are a practical example of TFL that is based on the concept of a doubly constrained, spatial interaction model where trip numbers are directly proportional to the product of the number of commuters in a production zone and the number of jobs in an attraction zone, and, most significantly, inversely proportional to some power of the distance between the two zones. Put simply, the shorter the distance between the production and attraction zones, ceteris paribus, the more commuter trips will result; all traffic analysis zones (TAZs) are related but near zones are more related than distant zones.

Spatial Interpolation and TFL

Spatial interpolation is the science of estimating the values of a surface at unsampled locations within a region based on a set of sampled values at various locations and is commonly found in GIS software such as ArcGIS and Idrisi. The value of a surface at an unsampled location is dependent of the values at nearby sampled locations. Spatial interpolation is a way to operationalize TFL. It assumes that the values of a surface exhibit some form of spatial autocorrelation.

Distance-weighted Averaging

A common method of spatial interpolation is to use a distance-weighted average. Decision variables in a distance-weighted average include the number of sampled values to incorporate, usually four, six, or eight. In Tobler's original words, this determines just how "parochial" or localized the estimating function should be (in kernel-based spatial interpolation the kernel can be limited by distance or number of points used) – the more localized, the more irregular the surface of estimated points. A second decision variable is whether there should be a spatial bias in choosing the nearest four, six, or eight points, that is whether they should be the nearest point in each one of four quadrants, six sextants, or eight octants, respectively, the object being to reduce directional bias in the estimation. The distance-weighting function may be exponential or simply $1/d$ where d is the distance from the unsampled point or $1/d^n$ where n is some power of distance. This allows "near things to be more related than distant things". The degree to which a distance-weighted interpolation is parochial depends on both the number of points included in the averaging process and the distance function used. In theory, all sampled points might be included (where everything is related to everything else) and this would produce a relatively smooth surface, the degree of smoothness being a function of how quickly the influence of distance declined. Trend surface analysis uses a global polynomial with two independent locational variables (X and Y; easting and northing; longitude and latitude) to estimate values of the dependent variable using a single equation. The polynomial can be first, second, or third order, or some higher number producing surfaces with additional inflections each time the order increases. Tobler referenced Richard Chorley and Peter Haggett's early use of such trend surface models.

Kriging

Kriging is another commonly used method of spatial interpolation. Kriging allows the researcher more objective control over the variables implicit in TFL, more objective control, that is, than the subjective choices used in distance-weighted averaging. Specifically, in kriging everything is not "related to everything else" because the "relatedness" ceases when the range is reached. The range is a distance over which points no longer covary. The extent to which points covary is measured by the covariance function that is estimated by a variogram. The variogram can take a large variety of functional forms, and this is similar to the changing influence of distance in a distance-weighted averaging. The variogram is an estimation of just how strongly "near things" are related. It also shows how quickly "relatedness" declines to zero. Universal kriging (as opposed to simple kriging) allows the geographic researcher to model the influence of spatial trends in the data since "near things" in one direction may be "more related" than near things in another direction. For example, data in a north-south direction might be more highly correlated than in an east-west direction.

In many instances, near things may be related not only spatially but through the influence of one or more variables. Thus, population density in a city such as Detroit might be influenced by many variables. Yeates found that land values at any given location in Chicago were influenced by distance from the central business district, distance from the nearest regional shopping center, distance from Lake Michigan, distance from the nearest elevated subway system, population density, and percentage of nonwhite residents at the given location. The land values at any location might be "related" to the land values at nearby locations, as TFL states, but this relationship would be influenced by these other variables and would therefore be better modeled by co-kriging than by simple kriging or regression as suggested by Yeates.

First Laws in other Disciplines

The idea of a first law may be found in other disciplines, often with partially humorous intent. Examples abound on the Internet: "The first law of Politics is that it is always about politics"; in a similar vein, the first law of sociology might be given as: "Some do, some don't" but this statement is usually attributed apocryphally to the physicist, Ernest Rutherford, as an insult to social science. Weisburd has identified "a first law of the criminology of place—the law of crime concentration at places" but this seems to be simply a special case of TFL as is Sociology's homophily and perhaps as are many other examples from disciplines that study spatially distributed data. Science provides some rigorous, universally accepted examples, such as the first law of thermodynamics. Geologists have given the name "uniformitarianism" to the idea that the natural laws governing geological processes that operate today also operated throughout geological history. More wittily, one researcher has suggested that the first law of geology should be that "the rocks remember while liquids and gases forget". Since kriging and, to a large extent, geostatistics were developed in geology, it is not surprising that TFL has recently been claimed by geologists, without acknowledgement: "According to the first law of geology 'everything is related to everything, but near things are more related than distant things'" Even traditional sciences such as biology find law-making difficult although it has been suggested that biology's first law should be that there is a tendency for diversity and complexity to increase in evolutionary systems. It is interesting to note that evolution itself is hindered by TFL, since a primary method of speciation, allopatric speciation, requires the isolation of two breeding populations by geographical barriers (mountains, rivers, oceans), migration, or habitat fragmentation. This is a process known as geographic vicariance. It is a process that can be reversed through genetic admixture if a geographic barrier is removed by, for example, a river ceasing to flow. Then TFL would re-establish itself. Perhaps TFL should be classified as a dangerous idea, since, like globalization, it inhibits diversity and complexity. This would make the inverse of TFL, spatialization, which Montello et al. claim is the first law of cognitive geography, namely "that people believe closer things to be more similar than distant things," attractive only for human information interface design and not for designing housing communities and cities such as Detroit.

Tobler himself cites several "first laws" from other disciplines (some closely allied to geography and some not) including Ernst Georg Ravenstein's well-known laws of migration and George Zipf's law, which stated that the frequency of a word in a large corpus of text would be inversely proportional to its frequency rank. This is also known as a power law or the rank-size rule, the latter suggesting perhaps more reasonably a diminished status. Power laws have been applied extensively to the population ranks of cities in numerous countries and to the study of social networks in geography to detect scale invariance but, unfortunately, they can be derived in many different ways and therefore tell us little about generative processes. Within subdisciplines of geography, or those closely related either as historical spinoffs or cognate subject matter, there are many examples of law-like statements although few "first laws". For example, in fluvial geomorphology Robert Horton's laws of stream numbers and stream lengths and in biogeography the work of Robert MacArthur and Edward Wilson developing a theory of island biogeography that linked species richness to an island's area and isolation. In the subdisciplines of human geography, as might be expected, there are fewer attempts to identify law-like behavior although the first law of cognitive geography, mentioned above, is a notable exception. It might be expected that GIScience would have made progress in identifying this new subdiscipline of Geography. Reitsma argues that many of the "laws" of Geography such as TFL can not be claimed for GIScience but they do guide how research in this nascent subdiscipline should be conducted. Nevertheless, the prognosis for new laws of GIScience is good since the subdiscipline has all the epistemic values that are considered necessary for scientific research: "simplicity, predictive accuracy, fertility, coherence with existing knowledge, unification and testability".

Spaces where Tobler's First Law Fails

Borders, Boundaries and Barriers

TFL will fail (or provide an incomplete explanation) wherever there are natural barriers similar to those that have aided the process of evolution. It will also fail wherever there are political or administrative barriers. A striking example may be found in the land use on the border between some parts of Haiti and the Dominican Republic, on the island of Hispaniola. Even the land use north and south of the forty-ninth parallel of latitude, forming the border between the United States and Canada, may show dramatic differences. In many countries, census divisions and political constituencies have boundaries that are deliberately constructed to maximize internal homogeneity and external heterogeneity in terms of socioeconomic characteristics and voting behavior. Ordinarily, having a political unit composed partly of an urban area and partly of a rural area would be regarded as gerrymandering. Finally, some countries may erect barriers between themselves and neighboring countries. Examples include the former Berlin Wall separating East and West Berlin, the Mexico-US barrier or border fence, and the Israeli West Bank and other separation barriers. The existence and impact of such separation barriers have been prominent in fiction, including the celebrated work of Gloria Naylor who, in her first and second novels respectively, located the African American communities of Brewster Place and Linden Hills bordering each other but separated by a wall. "Linden Hills is a posh upper-middle-class settlement, Brewster Place the last stop on the road to the bottom in American society with a great distance between them even as they are both black". TFL stops at the wall in both the real and the fictional world. TFL does not readily explain ethnic segregation. Having noted these examples at microgeographical scales, it is important to state that some researchers have found evidence suggesting "that national cultures are organized geographically; similar cultures are found in neighbouring countries, and dissimilar cultures in countries that are far apart".

Hierarchical Spaces and Networks

Diffusion and migration processes often progress hierarchically. A diffusion of an innovation may be passed and adopted from one individual of influence to another. Once these individuals have adopted the innovation it may then be adopted by a process of filtering down through the hierarchy of their networks of influence. Such hierarchical diffusion may also occur across space. An innovation may be adopted by higher-order centers in a central place hierarchy and may then, in a second wave of adoptions, be embraced by individuals in lower-order centers. Indeed, the whole adoption process might filter down through the entire urban system in a hierarchical fashion. Geographical research has shown that the temporal spread of diseases such as severe acute respiratory syndrome (SARS), which, in 2003, spread from its origins in China to 25 other countries and to Taiwan, is primarily due to airline network accessibility. Bowen and Laroe estimated the parameters of a multiple regression model where the dependent variable was the number of days from the first case of the disease in a newly SARS-infected country (SIC) until the disease was contained on July 5, 2003. The higher the number of days, the earlier the outbreak, that is to say, the more quickly the disease moved from China, its origin, to the country concerned. Six independent variables were included in the regression model: the number of Chinese ancestry residents of a country as a percentage of the total population; a binary variable that was assigned a 1 for the 11 countries with the highest foreign direct investment in China and a 0 for the remaining countries; per capita gross national income; a country's 2003 population; a measure of accessibility from Beijing using airline network schedules; and distance between the SIC and China. Only the airline network accessibility and the number of Chinese ancestry residents as a percent of the total population were statistically significant. Distance itself was not, and so it can be argued that TFL did not apply to the diffusion of the 2003 SARS outbreak unless we measure "distance" as a function of network accessibility. In addition, the medical geographer would need to take past migration patterns into consideration as well. Bowen and Laroe cite other examples of disease propagation via transportation routes, including the transmission of the fourteenth-century outbreak of bubonic plague along sea trade routes and three outbreaks of cholera in the United States during the middle of the nineteenth century along existing trade routes. A county not on a trade route was not infected, regardless of "nearness".

Social Media and Social Networks

Social media (SM) and social networks (SNs) have weakened the influence of distance, contributing to the so-called death of distance and the weakening of TFL. Waters has reviewed both the literature on the death of distance, including Cairncross's book on this topic, and the research on the impact of SM and SNs on distance-decay functions within social networks. The primary attraction of SM and SNs, and even the Internet itself, has been to weaken or eliminate the influence of geography and distance and, if this is true, Waters has argued that a second law of geography might replace the first: "Everything is connected to everything else, but things more closely connected [in network space] are more related – and geography may well be irrelevant". In SN space there is also a problem with defining what is local. Christakis and Fowler have discussed this issue and describe research that demonstrates that it is our friends', and friends of friends', and friends of friends of friends' behavior that affects us, and it is these linkages that determine whether we are obese, happy, or smoke, or have heart disease, and presumably other characteristics as well. We may be connected to everyone else in the network, but influence ceases to be significant with three degrees of separation. Sociologists have identified homophily as "the principle that a contact between similar people occurs at a higher rate than among dissimilar people". Although their observation relates to social networks they argue that Geography is the most important of several influential independent variables.

However, the question remains as to whether distance is now irrelevant. Or perhaps more alliteratively: "Is spatial special when it's social?". The existing research (which is expanding rapidly as SN companies such as Facebook and LinkedIn utilize their own research teams) implies that the more "social" as opposed to "professional" the SN is the more distance still matters. Tobler himself suggested a second law that "the phenomenon external to a geographicarea of interest affects what goes on in the inside" – essentially a problem of system closure presumably because "everything is related to everything else".

Various researchers have fitted distance-decay functions to "friends" and connections within SNs. For example, the probability that two cell phone users are connected is proportional to the Euclidean distance separating them raised to the power -2. This is the classic distance-decay function used in a gravity model and is reminiscent of the findings of Mackay . Mackay's seminal study of long-distance phone calls between cities in the Province of Quebec, Canada, showed a distance-decay exponent of -0.9, but between cities in the Province of Ontario it was -1.7, a steeper distance-decay function. Rapid drop-offs in call interactions occurred between Quebec-Ontario city pairs in which case the telephone traffic was about one-fifth what it was between pairs of cities in Quebec. An even more dramatic drop-off was noted between Quebec-US city pairs, where the traffic was about one-fiftieth of the Quebec pairs. Such differences presumably occurred because there were fewer contacts across provincial and national borders and because of higher long-distance charges. Today cell phone rates may also influence such long-distance interactions and cross-jurisdictional calls. These provincial and national borders do not prevent interaction; rather, they modify the process making it impossible to use a single function to model it. Waters notes that one study of the SN, LiveJournal, determined that the average user had eight friends and that, again on average, 5.5 were "geographically influenced" and lived nearby with a distance-decay function where the exponent of distance was approximately -1, while the remaining 2.5 friends were a result of nongeographic processes, the former geographically influenced connections being a partial vindication of TFL. Other research cited by Waters also suggests that SNs are influenced by a mix of geographic and social influences. Such research has suggested that "node locality" or network-based metrics be combined with "geographic clustering metrics" (and presumably socioeconomic factors similar to the past migration patterns elucidated by Bowen and Laroe) to provide a modeling framework with exceptionally high explanatory power but lower generality than Tobler's Detroit model.

Tobler: The Last Word on Tobler's First Law

Daniel Sui, the convener of the 2003 AAG panel session on TFL, asked Waldo Tobler to review and respond to the participants' written reviews of his law. Tobler introduced his comments by saying that the goal of his model and of TFL was to emphasize simplicity and therefore generality. This allowed Tobler to predict population growth in Detroit, a modest goal but a goal that is most useful in transportation planning and for urban budgeting wherever these budgets are based on population numbers. Tobler states that Barnes was correct in observing that he was "restricting himself to local effects". Unfortunately, this is not what the law states when it declares "everything is related to everything else, but near things are more related than distant things," implying a gradual decline in relatedness with no sharp break and no end to the relatedness. However, spatial interpolation methods such as kriging, with its concept of the range, spatial smoothing's use of kernels and geographically weighted regressions, discriminant analysis, and principal component analysis all allow analysis to be restricted to the local. Tobler confines his subsequent comments to three issues: (i) What defines a law and does TFL satisfy the definition? (ii) Is it true "everything is related"? (iii) How should "near" be defined?

GEOARCHAEOLOGY

Geoarchaeology is a multi-disciplinary approach which uses the techniques and subject matter of geography, geology, geophysics and other Earth sciences to examine topics which inform archaeological knowledge and thought. Geoarchaeologists study the natural physical processes that affect archaeological sites such as geomorphology, the formation of sites through geological processes and the effects on buried sites and artifacts post-deposition. Geoarchaeologists' work frequently involves studying soil and sediments as well as other geographical concepts to contribute an archaeological study. Geoarchaeologists may also use computer cartography, geographic information systems (GIS) and digital elevation models (DEM) in combination with disciplines from human and social sciences and earth sciences. Geoarchaeology is important to society because it informs archaeologists about the geomorphology of the soil, sediments and the rocks on the buried sites and artifacts they're researching on. By doing this we are able to locate ancient cities and artifacts and estimate by the quality of soil how "prehistoric" they really are.

Techniques Used

Column Sampling

Column sampling is a technique of collecting samples from a section for analyzing and detecting the buried processes down the profile of the section. Narrow metal tins are hammered into the section in a series to collect the complete profile for study. If more than one tin is needed they are arranged offset and overlapping to one side so the complete profile can be rebuilt offsite in laboratory conditions.

Loss on Ignition Testing

Loss on ignition testing for soil organic content – a technique of measuring organic content in soil samples. Samples taken from a known place in the profile collected by column sampling are weighed then placed in a fierce oven which burns off the organic content. The resulting cooked sample is weighed again and the resulting loss in weight is an indicator of organic content in the profile at a certain depth. These readings are often used to detect buried soil horizons. A buried soil's horizons may not be visible in section and this horizon is an indicator of possible occupation levels. Ancient land surfaces especially from the prehistoric era can be difficult to discern so this technique is useful for evaluating an area's potential for prehistoric surfaces and archaeological evidence. Comparative measurements down the profile are made and a sudden rise in organic content at some point in the profile combined with other indicators is strong evidence for buried surfaces.

Near-surface Geophysical Prospection

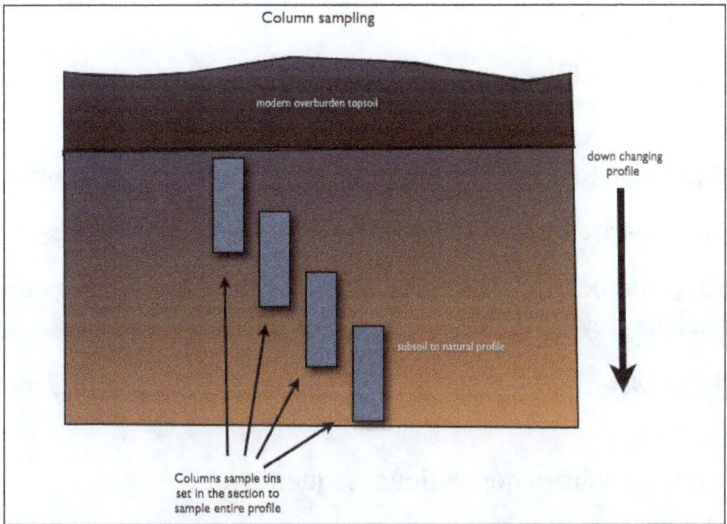

Offset column sampling of the soil profile.

Geophysical archaeological prospection methods are used to non-destructively explore and investigate possible structures of archaeological interest buried in the subsurface. Commonly used methods are:

- Magnetometry.

- Ground-penetrating radar.

- Earth resistance measurements.

- Electromagnetic induction measurements (including metal detection and magnetic susceptibility surveys).

- Sonar (sidescan, single-beam or multibeam sonar, sediment sonar) in underwater archaeology.

Less commonly used geophysical archaeological prospection methods are:

- Reflection or refraction seismic measurements.

- Gravity measurements.

- Thermography.

Magnetic Susceptibility Analysis

The magnetic susceptibility of a material is a measure of its ability to become magnetised by an external magnetic field. The magnetic susceptibility of a soil reflects the presence of magnetic iron-oxide minerals such as maghaematite; just because a soil contains a lot of iron does not mean that it will have high magnetic susceptibility. Magnetic forms of iron can be formed by burning and microbial activity such as occurs in top soils and some anaerobic deposits. Magnetic iron compounds can also be found in igneous and metamorphic rocks.

The relationship between iron and burning means that magnetic susceptibility is often used for:

- Site prospection, to identify areas of archaeological potential prior to excavation.

- Identifying hearth areas and the presence of burning residues in deposits.

- Explaining whether areas of reddening are due to burning or other natural processes such as gleying (waterlogging).

The relationship between soil formation and magnetic susceptibility means that it can also be used to:

- Identify buried soils in depositional sequences.

- Identify redeposited soil materials in peat, lake sediments etc.

Phosphate and Orthophosphate Content with Spectrophotometry

Phosphate in man-made soils derives from people, their animals, rubbish and bones. 100 people excrete about 62 kg of phosphate annually, with about the same from their rubbish. Their animals excrete even more. A human body contains about 650 g of PO_4 (500 g–80% in the skeleton), which results in elevated levels in burial sites. Most is quickly immobilised on the clay of the soil and 'fixed', where it can persist for thousands of years. For a 1 ha site this corresponds to about 150 kg PO_4 ha-1yr-1 about 0.5% to 10% of that already present in most soils. Therefore, it doesn't take long for human occupation to make orders of magnitude differences to the phosphate concentration in soil. Phosphorus exist in different 'pools' in the soil 1) organic (available), 2) occluded (adsorbed), 3) bound (chemically bound). Each of these pools can be extracted using progressively more aggressive chemicals. Some workers (Eidt especially), think that the ratios between these pools can give information about past land use, and perhaps even dating.

Whatever the method of getting the phosphorus from the soil into solution, the method of detecting it is usually the same. This uses the 'molybdate blue' reaction, where the depth of the colour is proportional to phosphorus concentration. In the lab, this is measured using a colorimeter, where light shining through a standard cell produces an electric current proportional to the light attenuation. In the field, the same reaction is used on detector sticks, which are compared to a colour chart.

Phosphate concentrations can be plotted on archaeological plans to show former activity areas, and is also used to prospect for sites in the wider landscape.

Particle Size Analysis

The particle size distribution of a soil sample may indicate the conditions under which the strata or sediment were deposited. Particle sizes are generally separated by means of dry or wet sieving (coarse samples such as till, gravel and sands, sometimes coarser silts) or by measuring the changes of the density of a dispersed solution (in sodium pyrophosphate, for example))of the sample (finer silts, clays). A rotating clock-glass with a very fine-grained dispersed sample under a heat lamp is useful in separating particles.

The results are plotted on curves which can be analyzed with statistical methods for particle distribution and other parameters.

The fractions received can be further investigated for cultural indicators, macro- and microfossils and other interesting features, so particle size analysis is in fact the first thing to do when handling these samples.

Trace Element Geochemistry

Trace element geochemistry is the study of the abundances of elements in geological materials that do not occur in a large quantity in these materials. Because these trace elements' concentrations are determined by a large number of particular situations under which a certain geological material is formed, they are usually unique between two locations which contain the same type of rock or other geological material.

Geoarchaeologists use this uniqueness in trace element geochemistry to trace ancient patterns of resource-acquisition and trade. For example, researchers can look at the trace element composition of obsidian artifacts in order to "fingerprint" those artifacts. They can then study the trace element composition of obsidian outcrops in order to determine the original source of the raw material used to make the artifact.

Clay Mineralogy Analysis

Geoarchaeologists study the mineralogical characteristics of pots through macroscopic and microscopic analyses. They can use these characteristics to understand the various manufacturing techniques used to make the pots, and through this, to know which production centers likely made these pots. They can also use the mineralogy to trace the raw materials used to make the pots to specific clay deposits.

Ostracod Analysis

Naturally occurring Ostracods in freshwater bodies are impacted by changes in salinity and pH due to human activities. Analysis of Ostracod shells in sediment columns show the changes brought about by farming and habitation activities. This record can be correlated with age dating techniques to help identify changes in human habitation patterns and population migrations.

Archaeological Geology

Archaeological geology is a term coined by Werner Kasig in 1980. It is a sub-field of geology which emphasises the value of earth constituents for human life.

GEOGRAPHICAL FEATURE

Geographical features are naturally-created features of the Earth. Natural geographical features consist of landforms and ecosystems. For example, terrain types, (physical factors of the environment) are natural geographical features. Conversely, human settlements or other engineered forms are considered types of artificial geographical features.

Natural Geographical Features

Ecosystems

There are two different terms to describe habitats: ecosystem and biome. An ecosystem is a community of organisms. In contrast, biomes occupy large areas of the globe and often encompass many different kinds of geographical features, including mountain ranges.

Biotic diversity within an ecosystem is the variability among living organisms from all sources, including inter alia, terrestrial, marine and other aquatic ecosystems. Living organisms are continually engaged in a set of relationships with every other element constituting the environment in which they exist, and ecosystem describes any situation where there is relationship between organisms and their environment.

Biomes represent large areas of ecologically similar communities of plants, animals, and soil organisms. Biomes are defined based on factors such as plant structures (such as trees, shrubs, and grasses), leaf types (such as broadleaf and needleleaf), plant spacing (forest, woodland, savanna), and climate. Unlike ecozones, biomes are not defined by genetic, taxonomic, or historical similarities. Biomes are often identified with particular patterns of ecological succession and climax vegetation.

Landforms

A landform comprises a geomorphological unit and is largely defined by its surface form and location in the landscape, as part of the terrain, and as such is typically an element of topography. Landforms are categorized by features such as elevation, slope, orientation, stratification, rock exposure, and soil type. They include berms, mounds, hills, cliffs, valleys, rivers, and numerous other elements. Oceans and continents are the highest-order landforms.

A body of water is any significant accumulation of water, usually covering the Earth. The term "body of water" most often refers to oceans, seas, and lakes, but it may also include smaller pools of water such as ponds, creeks or wetlands. Rivers, streams, canals, and other geographical features where water moves from one place to another are not always considered bodies of water, but they are included as geographical formations featuring water.

References

- Petit, Rémy J.; Aguinagalde, Itziar; Beaulieu, Jacques-Louis de; Bittkau, Christiane; Brewer, Simon; Cheddadi, Rachid; Ennos, Richard; Fineschi, Silvia; Grivet, Delphine (2003-06-06). "Glacial Refugia: Hotspots But Not Melting Pots of Genetic Diversity". Science. 300 (5625): 1563–1565. Bibcode:2003Sci...300.1563P. doi:10.1126/science.1083264. ISSN 0036-8075. PMID 12791991

- Sánchez-González, D.; Rodríguez-Rodríguez, V. (2016). Environmental Gerontology in Europe and Latin America. Policies and perspectives on environment and aging. New York: Springer Publishing Company. p. 306. ISBN 978-3-319-21418-4

- Dreesen, Roland, Dusar, M. and Doperé, F., 2001 . Atlas Natuursteen in Limburgse monumentenx- 2nd print 320pp. . LIKONA ISBN 90-74605-18-4

- Geo-literacy: geo.appstate.edu, Retrieved 15 March, 2020

Permissions

All chapters in this book are published with permission under the Creative Commons Attribution Share Alike License or equivalent. Every chapter published in this book has been scrutinized by our experts. Their significance has been extensively debated. The topics covered herein carry significant information for a comprehensive understanding. They may even be implemented as practical applications or may be referred to as a beginning point for further studies.

We would like to thank the editorial team for lending their expertise to make the book truly unique. They have played a crucial role in the development of this book. Without their invaluable contributions this book wouldn't have been possible. They have made vital efforts to compile up to date information on the varied aspects of this subject to make this book a valuable addition to the collection of many professionals and students.

This book was conceptualized with the vision of imparting up-to-date and integrated information in this field. To ensure the same, a matchless editorial board was set up. Every individual on the board went through rigorous rounds of assessment to prove their worth. After which they invested a large part of their time researching and compiling the most relevant data for our readers.

The editorial board has been involved in producing this book since its inception. They have spent rigorous hours researching and exploring the diverse topics which have resulted in the successful publishing of this book. They have passed on their knowledge of decades through this book. To expedite this challenging task, the publisher supported the team at every step. A small team of assistant editors was also appointed to further simplify the editing procedure and attain best results for the readers.

Apart from the editorial board, the designing team has also invested a significant amount of their time in understanding the subject and creating the most relevant covers. They scrutinized every image to scout for the most suitable representation of the subject and create an appropriate cover for the book.

The publishing team has been an ardent support to the editorial, designing and production team. Their endless efforts to recruit the best for this project, has resulted in the accomplishment of this book. They are a veteran in the field of academics and their pool of knowledge is as vast as their experience in printing. Their expertise and guidance has proved useful at every step. Their uncompromising quality standards have made this book an exceptional effort. Their encouragement from time to time has been an inspiration for everyone.

The publisher and the editorial board hope that this book will prove to be a valuable piece of knowledge for students, practitioners and scholars across the globe.

Index

www.ingramcontent.com/pod-product-compliance
Lightning Source LLC
Chambersburg PA
CBHW080402190526
45161CB00003B/107